The Mexican Urban Household

The Mexican Urban Household

Organizing for Self-Defense

by
Henry A. Selby,
Arthur D. Murphy,
and Stephen A. Lorenzen
with
Ignacio Cabrera,
Aida Castañeda,
and Ignacio Ruiz Love

UNIVERSITY OF TEXAS PRESS, AUSTIN

First Edition, 1990

Requests for permission to reproduce material from
this work should be sent to Permissions, University
of Texas Press, Box 7819, Austin, Texas 78713-7819.

∞ The paper used in this publication meets the
minimum requirements of American National
Standard for Information Sciences—Permanence of
Paper for Printed Library Materials, ANSI Z39.48-1984.

Library of Congress Cataloging-in-Publication Data
Selby, Henry A.
    The Mexican urban household : organizing for
self-defense / by Henry A. Selby, Arthur D. Murphy,
and Stephen A. Lorenzen ; with Ignacio Cabrera,
Aida Castañeda, and Ignacio Ruiz Love. — 1st ed.
        p.   cm.
    Includes bibliographical references.
    ISBN 0-292-78521-6 (alk. paper)
    1. Cost and standard of living—Mex-
ico. 2. Household surveys—Mexico. 3. Work
and family—Mexico. 4. Urban poor—Mexico. 5. Mexico—Economic conditions—
1982– . I. Murphy, Arthur D. II. Lorenzen,
Stephen A. (Stephen Alfred), 1946– . III. Title.
HD6996.S45 1990
306.85'0972'091732—dc20   90-31487
                                                    CIP

Contents

Preface

This book reports on qualitative and quantitative research by a team of anthropologists-sociologists-architects-economists. The book becomes more quantitative as it proceeds and, in that sense, more "economistic." The ideas are shared, so that, as in all good projects, we do not know whose ideas they originally were. We all share the anthropological bias of tending to see things from the point of view of labor rather than of capital, and the feeling of mission to do what we can to help ameliorate the lives of the poorer people in Mexican society. But none of us is a revolutionary. We think that academics who counsel revolution are no better than coat-holders, encouraging others to fight while staying on the sidelines themselves. And we feel, along with many who know Mexico, that once the machetes come out, it is a long time before they get sheathed again.

Many people deserve our thanks. Our greatest debt must be to our *patrón*, Ing. José Luis Aceves, who not only supported our research for ten years, but also restrained his criticism while viewing with alarm the rather spontaneous and irregular methods of the anthropologists, compared to the engineers. Arq. José María Gutiérrez was an intellectual mentor, as well as a critical influence and, when it looked as though the project was going to fail, showed us how really to operate in the Mexican political scene. Bill Glade, the director of the Institute of Latin American Studies at the University of Texas at Austin during most of the years of the project was sympathetic, helpful, and always supportive. Mike Conroy, the current associate director of the same institute, was both stimulating colleague and generous friend throughout.

Thanks are due to the institutions that supported us through the years of the project. The National Science Foundation has been generous in its support, especially for the work in Oaxaca. The Fulbright-Hays program made it possible for Arthur Murphy to work with the Instituto Nacional para el Desarrollo de la Comunidad y de la Vivienda Popular (INDECO) during 1976–77. Stephen Lorenzen would like to thank the Tinker Foundation for supporting his work on small businesses in Mexico. Additional support was provided by the Rusk Center of the University of Georgia, the University Research Committee at Baylor University, the Office of Urban Studies of the AID, the Department of State, and the Institute of Latin American Studies at the University of Texas. The major funding came from INDECO-Mexico, which carried out the survey work on the 10-city sample.

It is hard to thank the people who really deserve it, the Mexican householders themselves. The project started with such high hopes for Mexico; as the president said, Mexico was preparing itself for the delicious and hard-won task of "administering prosperity." But at the time of writing it is in the seventh year of the worst depression since the Revolution and the working people of Mexico have lost over 40 percent of the purchasing power of their wages. As one thoroughly piqued and disgruntled laid-off carpenter said to us, "I complain that there is never any meat to eat, but my kids tell me that we are all vegetarians now." Working Mexico has moved from the prospect of prosperity to obligatory vegetarianism. Henry Selby owes a particular thanks to three Mexican benefactors: Aurelio Bautista of San Juan de Dios in Oaxaca; his compadre, Rosalino Galindo, and all his family who took him in in 1966. Lorensen and Selby want to thank a very enterprising and generous family in Ciudad Netzahualcoyotl who took them in and looked after them and their colleagues, students, and visiting firemen. The family asked to remain anonymous, but their gratitude remains. They also want to thank Myung-Hye Kim, who was both splendid colleague and enterprising researcher, who made sure that they got out the front door and did field work. Liliana Valenzuela, we thank for her patience with foreigners and her interviewing skill on the project.

Murphy would like to express his gratitude to the people of the city of Oaxaca for hosting him as he learned about their city and worked with INDECO to develop the survey method and instrument used in the 10-city sample. He owes special thanks to the people of San Juan Chapultepec and Riberas del Atoyac. They took a neophyte researcher and transformed him into an anthropologist. Thanks must go to Doña Victoria and Don Beto, who introduced him to life in the *colonias* of Oaxaca, and the families Bolaños and Pacheco, who have supported him, his colleagues, and students over the years.

Apart from our benefactors and patrons, our colleagues deserve our thanks and acknowledgment. First, to a *terna de norteños*, three constant companions and co-workers, go special thanks. Ignacio Ruiz Love from the Universidad Iberoamericana, the INDECO, and now a successful Sinaloan academic and entrepreneur, more than constant companionship is owed, for he insistently invoked the best in Latin American social science for the gringos and good-naturedly insisted on it far into the night. Ignacio Cabrera was always a calming influence, for he always seemed to understand the Mexican bureaucracy and political system better than anyone else. Now the former state delegate (administrator) for the Ministry of Urban Development and Ecology in the state of Sonora, he has a very bright political future in Mexico and is responsible, as is none of the rest of us, for implementing the prescriptions that follow from this, his study. Aida Castañeda was indispensable not just as friend, translator, and hard-working colleague who supervised the field work and data collection in the largest sample of all in the city of Reynosa, but she also directed us to keep our sights on comparative material and to look outside the Eurocentric-American capitalist world to see larger perspectives. It is with pleasure and justice that we include them on the title page as associates in this enterprise.

We want to thank our families who have put up with frequent absences, unannounced visitors, demands, and accidental (in the Spanish-language sense) emotional states. It is difficult, sometimes, to separate life from project, and we deeply thank those that love us and forbear to ask why.

To Doug Uzzell and Jim Greenberg, special thanks for reading an earlier version and being so warmly encouraging and helpfully critical at the same time. Guillermo de la Peña, Mercedes González de la Rocha, and Salomón Nahmad, all of Centro de Investigaciones y Estudios Superiores en Antropología Social (CIESAS) have made invaluable comments. Larissa Lomnitz read the manuscript in raw form and therefore deserves our special thanks. The anonymous reviewers of the University of Texas Press were also excellent. Our editor Theresa May has been helpful ever since she was brought into the project, and we thank her too. Also thanks for a particularly professional copy-editing job by Barbara Cummings, who truly understands the rules for accenting Spanish and who saved us from too many errors. But perhaps the most helpful commentator of all has been Harley Browning, who insisted all the way through the project that if we did not do ethnography we would cheat our own talents. We have tried to live up to his encouragements.

Lastly, we would like to thank Carole Smith and Susan Lane of the University of Texas Department of Anthropology, who patiently typed the most difficult part of the manuscript, as well as Paulette Edwards and Jeanie Fitzpatrick of Baylor University's Department of Sociology, Anthropology, and Gerontology. When we were so sick of formatting tables that we were about to strangle each other, they took over and produced the whole manuscript.

The Mexican Urban Household

PART ONE

The Setting, the Study, and the Cities

In the first two chapters of the book, we introduce the study and the cities that were studied. Because there were so many kinds of "us" engaged in the study, it is necessarily diffuse in focus. We did not try to cover every base that should be covered in the various disciplines among us. We did try to encompass the urban experience and urban "reality" in such a way that the results of the study would be generalized and listened to by people whose lives are attuned to big-sample statistics. The major focus is anthropological: we discern the cities of Mexico through the ideological lens of the working class, the ordinary people of Mexico.

When we turn to study the 10 cities of the sample, we find that the urban experience of the majority of Mexicans is a matter of grays. Talking with license in the manner of a statistician, the most impressive fact about Mexican urban life is how the variance has been squeezed out of it. There are only the tiniest of differences between lives in such apparently disparate places as Mexico City and Oaxaca. Early on in the study, we did a quick and dirty cluster analysis based on our consumption data, and we found that there were two kinds of cities: "Oaxaca-type" cities and "Tampico-type" cities. Oaxaca-type places did not have a powerful oil union, while Tampico-type did. Nine of the 10 cities were Oaxaca-type, and only Tampico was Tampico. And only in Tampico did anything close to a large proportion of the ordinary households enjoy some kind of "modern," upscale consumption style. And even in Tampico, they were a minority. We conclude that Mexican urban life is mostly shades of gray.

But some differences can be discerned, and so in chapter 2 we try to distinguish between "good cities to live in" and "bad cities to live in," imploring the reader to remember that these are relative terms. We discover some interesting things: that San Luis Potosí is the best city for ordinary households to live in and that Oaxaca is the worst. In San Luis there is some chance at a decent job and a half-way decent salary, or there was during the precritical years when we collected the first tranche of data. There is not in Oaxaca. Charming, sleepy, colonial Oaxaca is hell to live in unless you happen to be a tourist, a bureaucrat, or an oligarch. The key, not surprisingly, is the labor market and, specifically, the availability of jobs in the most dynamic areas of the manufacturing sector. Urban infrastructure is also important, as is the availability of inexpensive housing and security of tenure, as well as the opportunity to form large complex (usually extended) households. The view of the big city from the *colonia popular* is decidedly practical and has mostly to do with feeding the family and keeping a place to live.

This depressing introduction to the rest of the study provides the *puesta en escena* for the rest of the study, which is a description and analysis of how the Mexican urban household organizes to defend itself, the subject of Parts Two and Three.

1. The Setting and the Study

Sra. Concepción Hernández[1] is a remarkable woman. She was abandoned by her husband after he had fathered five children, and now (as of July 1987) she lives in a new *colonia popular*, or poor neighborhood, on the outskirts of Oaxaca in a concrete house that fronts on the street, has a patio where she keeps a couple of chickens and a pig, and also boasts a well. She is a laundress. Her five children, the oldest of whom is 15, are all launderers, and the older two go about the neighborhood and into the city to collect the laundry and bring it home before they go off to school in the morning, and then they deliver it at night. If you walk into the patio of an afternoon from Monday to Saturday, the scene is touching and drenching: everyone is working. The children are excused from work only if they have homework, or if they have to raise money for the school through the sale of *boletos*, or raffle tickets, which is the way in which the school raises funds for basic expenses in an era of economic crisis.

To accomplish all this they get up around 5:30 in the morning and have breakfast (eggs, beans, *salsa*, tortillas) from around 7:00 to 8:00. At 8:15 the washing starts, with Sra. Hernández in charge. It goes on until 12:00, at which point she breaks off to make the main meal of the day, which usually resembles breakfast, since meat is very expensive and eggs will go further. At 3:30 the family wash starts, and the younger kids clean house and straighten up the patio. They do not have supper, but normally they do spend the evenings together. "At night," she says, "I like to read to the kids from the Bible, and then we have a couple of blessings, and then we go to bed." On Saturday the 13-inch black-and-white TV is turned on so that the older boy can watch the Saturday afternoon movie, and sometimes, it is turned on on Sunday for the football. "I worry," she says, "for TV can become a vice."

Sra. Hernández attracted our attention because of her robust self-confidence in the future and her strong feeling that she had done something very difficult in raising her five children and her feeling that she had done it well. She does not feel sorry for herself for one instant. She sometimes has occasion to go to the *zócalo* in Oaxaca, where she sees tourists sitting outside drinking and eating and laughing and relaxing. She sees them, but does not envy them, for, as she says, "These people, the tourists, they have troubles just like me, even though they earn more than the minimum salary."

When we protested that many a tourist would give a lot to have as wonderful a family as she had, she assented smiling, and said, by way of giving a subtitle for this book: "Con estos niños, voy a superar," or "Yes, and with these kids I am going to come out on top!" Her oldest son had just completed secondary school, or nine years of formal education. Maybe he could get the best of all possible jobs, a state job (no work, high pay, with all fringe benefits), failing that perhaps one of the wonderful jobs in the tourist industry ("It's wonderful the way those people leave tips right there on the table," she said in some awe), or failing that a job working as a helper in a body-work shop. He had no plans for leaving home and moving to Mexico City, not just because there were not any jobs there anymore, but because he was not going to abandon his mother and his family.

We asked Sra. Hernández, as we asked many Oaxaqueños in 1987, about whether they preferred small or large families. Her response was not much different from everyone else's, as she explained that small families were better on the whole because you did not have to suffer as much, and she added that large families could be very difficult if the parents did not look after the kids and the kids turned out to be *vagos* and got into drugs and other vices, but that was not true in her case. The main disadvantage to the small family was that the kids could not help each other. In the big family the older kids can work and the younger kids can study, but in the smaller family the kids have to study and work at the same time, and that is a big disadvantage if you are trying to get a good education. In a word, she assented to the government-sponsored notion that small families were better at the same time that she saw the advantages of the larger family in her front yard every day.

Concepción Hernández represents one way of rising out of the poverty that is the fate of all ordinary[2] Mexican families in the cities. By the force of her personality, mobilizing her children, she is able to establish and run a laundry firm made up of the members of her family. The firm is going to diversify as the children one by one leave secondary school and enter the job market, for there is no question in her mind nor in her children's that the first order of loyalty will be to the family (and to the mother) that raised them. There will be problems when the children come of age to marry, but those are far down the line, and with any luck a daughter-in-law can be carefully chosen who will want to join them and live with them in an extended family corporation.

Her comments about her situation incorporate the other themes of this book. If the family is first and foremost, the instrumentality of its own salvation, then the major constraint on its success is the labor market. People who work will always find it, as the saying has it, but some jobs are better than others, and the best jobs are those that are secure, from which you cannot get fired or laid off: state jobs. But she knows that the probability of her eldest getting one of these is small. The family does not have the connections, and he is just one kid among the thousands that would fight for one. Like most ordinary Mexicans, she does not think about the tediousness of the work, or the working conditions, or how boring some jobs are, since the problem is getting work, any work, and only the foolish would quibble about employment conditions if the pay was good and the work was steady.

Like most ordinary Mexicans, Sra. Hernández feels quite sure that the key to getting a decent job is education. She has no illusions about, or quarrels with, the content of that education, for it is the certificate that matters. She is perfectly aware that credentialism is on the increase ("You have to have a secondary school certificate now, whereas before all you needed was primary, if you want something other than a laboring job" [*trabajo sencillo*]). She knows that there is no guarantee that the certificate will deliver a job, but she knows that without a certificate, there is not a chance.

She recognizes, like most Mexican families, that things will improve with time, with changes in the domestic cycle and composition of her family, if she plays her cards right. She recognizes, like González de la Rocha's eloquent interviewee, Lupe (1986:98), "*es re'duro comenzar*," which means that it is very difficult indeed to set up a household, that those early years are really tough, more especially if your husband abandons you and you are left alone with small children, but what you have to do is care for your family first so that your family will care for you.

She knows that there are different ways of caring for the family. You can have a small one and invest heavily in the children, but as other people we interviewed said, "You are taking a chance. What happens if you have two children and they both go away and leave you on your own?" She recognizes that "if you have a small family you can give them what they need, what they want, and you don't have to yell at them all the time as when there are a lot of children." She recognizes that the investment that postsecondary education[3] involves in forgone wages, clothing, books, school supplies, and uniforms is very high and beyond her means, but that it is another riskier strategy for accomplishing what she is accomplishing: rising out of immiserated poverty to achieve a standard of living where one could say that the family is "getting by."

This book is about Concepción Hernández and about how the ordinary urban Mexican household organizes itself to get by in a difficult and challenging economic and social world.[4] We use the somewhat awkward word "ordinary" to describe the majority of Mexican households in the cities who are not upper nor middle class, but who constitute the majority in the country. Many of them are "working class" in the sense that their members work for a living sometimes in the labor force and sometimes not, but that term should not connote blue-collar work, for only 18 percent of the householders hold jobs like that (7% of the women and 21% of the men, as Kim, 1987:106, notes). The majority of Mexicans are poor and hardworking and always on the lookout for income-earning opportunities. Sometimes, among ourselves, we try to cut through the sectoral terminology by reminding ourselves that ordinary Mexicans have to hustle (but not in the criminal sense) for most of their lives. Though we have only fragmentary data on job histories, it is well to remind ourselves that today's factory worker is tomorrow's Kleenex salesman, often perforce but other times by desire. Ordinary Mexican workers are always looking for opportunities to better themselves and rightly regard our standard economic categories (even including such terms as "employed," "underemployed," and "unemployed") as being irrelevant to their ideas of what they go through to maintain their families. For Mexicans, more so than for Americans and more like the Europeans, work and jobs are means to an end; for the majority of Mexicans, that end is defending themselves and their family. People are judged more on how they do that job, rather than on whether they are good workers and able to keep the same jobs.

In this book we try to describe aspects of the domestic and working lives of urban Mexicans in a way that reflects their own perspectives. At the same time we try to present data and analysis on about 10,000 households to fit our own understanding.

First, we analyze the different cities that we studied to find out whether some are more livable than others from the point of view of the ordinary Mexican. This gives us a chance to talk about how cities appear to ordinary people, and how they evaluate them, and why some cities are better places to live in than others. From the point of view of ordinary households, the best city in our sample of 10 is San Luis Potosí, and not Mexico City because, at the time of the study, it provided the best jobs, best wages (in relation to the cost of living), and the best housing conditions for poor people. Charming, colonial Oaxaca was the worst: low wages, bad jobs, little dynamism in the economy, bad housing, and little prospect for improvement. A surprising finding is that the cities are more alike than they are different. For ordinary people there is a grayness to Mexican urban life in its material aspects, a remarkable homogeneity. Living in Venustiano Carranza, a working-class area of Mexico City, is very much like living in a working-class area in another city, despite the enormous apparent differences between the world's largest city and the provincial centers.

In the second part of the book (chaps. 3, 4, and 5), we discuss the Mexican urban household. In chapter 3 we provide a theoretical background to some of the more vexing questions having to do with the household. Is it a decision-making unit, properly speaking? What is the symbolic and moral organization of the households, and what contradictions does this moral ordering cause? Can we talk about the ideology of the household, as well as of family and kinship, and what are the daily uses of this ideology, and whose interests are favored? In particular is the ideology gender neutral? Finally, what is the ideal household in the Mexican popular ideology, and how closely do the statistical patterns that we observe conform to this pattern?

After this, in chapter 4, we provide a descriptive basis for the study of household and family by looking at major and minor household types, their incidence, and their economic and demographic characteristics. We find that singleton households are very rare and usually transitory and that there are three kinds of matrifocal families, only one of which conforms to the North American pattern so important in understanding the feminization of poverty, while the other two represent widows' households and/or the households of women who have apparently gone off on their own and done quite nicely without a senior male/husband about the house. We speculate that these may be cases where the bullying macho, or *pendenciero fanfarrón* in González de la Rocha's felicitous term, has had his comeuppance, been abandoned by his family, who are doing better without him than they were doing with him. And finally we look at the characteristics of the various kinds of complex households, both those with distant relatives present and those which have attained the ideal status of being linear extended families, about equally patrilineal and matrilineal.

In chapter 5 we present data to sustain the notion that the better-off households are larger; at the same time we show that the control of the worker dependency ratio[5] is all-important. Households in the earlier stages of the domestic cycle are in much poorer economic shape than households later in the cycle, but the oldest households are about in the same shape as the younger ones.

In Part Three (chaps. 6, 7, and 8) we look in some detail at the household economy. In chapter 6 we examine individual earnings in the household, in particular the earnings of people in the "formal" and "informal" sectors of the economy, and we find that under detailed examination, leaving out the value of the fringe benefits associated with formal sector employment, the differences are really quite small and that informal sector earning opportunities have advantages that might have gone unrecognized.

In chapter 7 we turn from the study of individual incomes to look at household budgets and incomes. We define per capita discretionary income as a measure of the potential flexibility in the household budget and look to see what households are better off in the sense of having more leeway in their budgets and more opportunities for investment in themselves and their activities. Now we begin to see the advantages of the smaller household. Sra. Hernández was quite correct. She told us that for overall welfare larger households were better, but that if the parents wanted to make investments in the children, smaller families were better. This leads us to suggest that the ideal household cum family arrangement would be a combination of these two strategies: large households made up of (a number of) small families. But the moral, emotional, and symbolic constraints on family living, particularly those deriving from the division of authority among men and women and budgetary difficulties seem to make this impossible except under conditions of exceptional economic stress. There are sexist and self-interested propositions on all sides of the issue concerning who is being robbed and who is the free rider in the extended family household. Mothers-in-law claim that their grasping daughters-in-law "counsel" their sons to favor his own family at the expense of the extended collective, while daughters-in-law complain that their husband's parents favor behavior like drinking and visiting, which is inimical to their interest and the interest of their small family. Daughters hear from their parents that their husbands do not contribute to the family budget nearly the amount that they should, and their husbands claim that her parents are heartless because they are unwilling to help and to do a favor when it is needed (i.e., in the early years of the marriage when it is re'duro to form a household). Keeping an extended household operating is a difficult proposition at the best of times and with the best of people. With more than one junior family, and in the worst of economic times, like the present, it is doubly difficult.

We go on to study household budgets and find that the elimination of housing expenses is the key to getting out of immiserated poverty. People who pay rent, as any "parachutist"[6] from the inner city will tell you, have next to no discretionary income. After reviewing the data on individual and household incomes and on household budgets, we are led to conclude that the implied strategy of the household is not the maximization of income, but rather the control of expenses.

In chapter 9 we look at a new data set gathered in Oaxaca in 1987, which allows us to see what has happened in one city after five years of a terrible economic crisis that has reduced individual wages more than 40 percent nationally and lowered household income 23 percent in Oaxaca. We find that the combined or compromise arrangement of small families living in large households is being instituted as a result of the economic crisis. We note that urban fertility is decreasing but not so much in Oaxaca, where the percentage of children under five years of age is the same as it was 10 years previously, but that households are larger (averaging 5.6 members in 1987 as compared to 5.2 members a decade earlier). We similarly note that, although employment opportunities for children have declined, women are entering the work force in unprecedented numbers, particularly married women, overwhelmingly in the informal sector.

This leads us to a final and speculative discussion about the "New Deal" for the Mexican majority, a policy that is being forced upon the ruling elements in Mexican society whether they know it or not. From the ordinary household's point of view, the bind they have been put in is intolerable. During the years of expansion before the crisis, it was possible to have a large number of children and then to mobilize them into a large household in order to survive in a minimally decent fashion. But the cost-benefit equation that underlies that arrangement has been upset by the changes during the crisis period. Now children cost more because they require secondary and sometimes postsecondary education to get a job for which an inexpensive primary education used to be sufficient. Children are still exceptionally useful around the house, performing all manner of unremunerated jobs that are necessary to the survival of the household, but their economic value in income terms has declined because of the decline in the availability of jobs for the young, particularly part-time jobs. Women, particularly married women, are taking their place in the work force. Now that the underlying economic rationale for large numbers of children has been eliminated on both the cost and benefit sides, we can expect that the "modernization" process of reducing family size can proceed apace. This process is already under way, so much so that it has offset the decrease in real wages and maintained the value of per capita discretionary income in the family and reduced the impact of the decline in household income by about 50 percent. Mexico is rapidly moving to the "American"-type household, with income-earning adults and dependent children, which requires a rather different type of (political) economy to sustain it. With time lags, it is also moving toward an American or more Scandinavian population pyramid in which there are proportionately fewer young entrants into the work force, constantly exerting downward pressure of wages. Right now, during the economic crisis, wages are falling, and there is no reason to believe that they will rise significantly for the next three years. At the same time the increased costs of preparing children for the work force is either going to have to be borne by the state, with an enormous increase in social spending, or it will have to be borne by the household, which has only the device of forming larger and larger collectives of smaller and smaller families to sustain it. Our feeling is that the process of household growth has reached a limit, although we must admit that we thought that the limit had been reached before now, and we acknowledge that the state has little money under the current distribution of income to spend on social services. So for the next few years, we have a Mexican work force that cannot upgrade its formal educational skills, while at the same time Mexico is trying to open its economy and enter into the General Agreement on Tariffs and Trade (GATT). Something has to give. Ordinary families cannot and will not make further sacrifices, it seems to us. If neither the state through redistribution and tax policies nor the household through its "policies" of mobilization and economies of scale in consumption can generate the funds necessary for the education of the young for the work place, there is no way that Mexico could survive its attempt to liberalize its economy and enter the world market, except under conditions of extreme dependency and industrial backwardness. In order to resume some kind of orderly growth, Mexico is going to have to face up to the challenges being posed by changes in the Mexican family, household, and labor force, as well as the challenges of foreigners entering their domestic markets and the drain of local capitals as a result of payments on the foreign debt. In the work force to come, fewer workers will mean rising wages, but if there is any connection between education and productivity, as most mainstream economists believe, we shall not see the concomitant increases in productivity to sustain the growth. We shall certainly not see the growth in the highly skilled sectors of the labor force that will permit the maturation of the Mexican economy.

Redistribution of income is necessary now, and no Mexican regime since the Revolution has shown much stomach for it, present company not excepted. *Perestroika* is out in the world, it would seem, and in Mexico's case it is coming from below, where it is already taking place. Something, we repeat, has to give; the ordinary family has sacrificed as much as it will.

The Study

The first impetus for this study of urban Mexico was provided by an encounter in Oaxaca in 1975 between one of the contributors (Murphy), his colleague, Alex Stepick, and Ing. José Luis Aceves, the Oaxaca state director of the National Institute of Community Development (INDECO) who asked why the inhabitants of a local neighborhood were not using the facilities of the clinic that had been built for them. Since both Murphy and Stepick had been living in the neighborhood, they knew why, and they were able to acquaint the engineer with the kinds of communication problems between the neighborhood and his institute that are so common everywhere.[7] Ing. Aceves, who was to become our *patrón* after that, was impressed by these North Americans who found out about poor neighborhoods by living there—found himself hooked by anthropology, in other words—and from that time on our association has grown. Aceves turned to Murphy for advice and help when the Mexican agency director was asked to carry out the pilot study for a national survey of the living conditions of the "unsalaried" householders of Mexico, and it is the subsequent national study that we report upon here. It was projected to cover all 73 Mexican cities with populations over 50,000 in the 1970 census. Such ambitiousness was typical of the early López Portillo years when, in the historically inaccurate phrase of the president, Mexico was to prepare itself "for the administration of prosperity." In the end we were able to study 10 cities, with a total of 9,458 households.

The study was a continuation of one that we had started in the 1970s, when Murphy, Selby, and Stepick spent a year studying the goals and strategies of Oaxacan villagers and city dwellers and trying to construct a quantified production model of them. Together with

Gary Hendrix we wrote a number of essays about ordinary people's goals and the ways in which they attempted to attain them in rural and urban Oaxaca (see, e.g., Murphy, 1973; Selby and Hendrix, 1976; Stepick, 1974; and Stepick and Hendrix, 1973). One of the most important results of the study was the connection forged between Stepick and Murphy and the (Mexican) National Institute of Community Development (INDECO) and its competent and incorruptible director Ing. José Luis Aceves.

The 1970s were the palmy days for planners in Mexico: it was still the second development decade, and the efficacy of planning was taken for granted as forward looking, necessary, and rational. Wags were not yet saying, as they later did, that there were three kinds of planning: *planificación socialista,* as practiced in the socialist countries; *planificación indicativa,* as practiced in France and Scandinavia; and *planificación decorativa*—the Mexican version. We felt that there would be funds for development, that there was a commitment to social justice in the party, the government, and the country, and that our task was to help funnel investment toward appropriate projects so as to make best use of available funds. We defined "best use" in the way that it had been defined in our earlier studies of urban life: the use that enabled city dwellers most efficiently to realize their goals as they defined them.

But this lay in the future. First, we had to discover what were the prevailing conditions of life. For this purpose Murphy, Ignacio Ruiz, Ignacio Cabrera, and Aida Castañeda in consultation with senior members of INDECO and Ing. Aceves adapted methods that had been pioneered by the School of Architecture of the Massachusetts Institute of Technology for urban studies and modified them to fit the Mexican case.

The methods are described in Appendix 1. For readers who may not be interested in the details, the methods can be briefly described. First, cities were "walked" by teams of social scientists, architects, and engineers, and a map was drawn of the 24 "localities" that made up the city, as defined by their geographic barriers (highways, rivers, *barrancas,* etc.). At the same time a detailed typology of neighborhood subtypes was defined, that is, areas where people had the same house style, lot tenure, services and amenities, income, and social advantages. Care was taken to make fine discriminations among neighborhoods at this initial stage. In Oaxaca, a city of less than 200,000 people in 1977, 116 neighborhood subtypes were defined. Two householders in each neighborhood subtype were interviewed about their living conditions and those of their neighbors, plans were drawn of their houses, and an inventory was made of the ameni-

ties to which they had access—from blenders to sewers to bus routes
to work. This data was used to construct a picture of the city, neighborhood by neighborhood, and the initial rough typology was re-
worked so as to incorporate data on their age, income, (roughly calculated from the interviews), house style (from bamboo shack to
substantial middle-class house with servants' quarters), and type of
house and/or lot tenure ("regular"/"irregular"). The 116 subtypes
were aggregated into eight neighborhood types: invasions; three
kinds of *colonia popular* according to the income level of the inhabitants; traditional towns that had been incorporated into the city
(*pueblos conurbados*); site-and-service projects (*auto-construcción*);
center city neighborhoods; two kinds of middle-class neighborhoods—planned and subsidized projects (Instituto Nacional para el
Fomento de Vivienda de los Trabajadores [INFONAVIT] and *conjun-
tos habitacionales*); and private developments and houses. The ty-
pologies changed from city to city, incorporating new types and dis-
carding irrelevant types, but the changes were not marked. One
representative of each (of the eight) neighborhood type(s) was taken
from each of the 24 (geographical) localities, yielding 52 subtype
sampling units. A quota sampling strategy was applied to the 52
sampling units so as to assure representativeness at the level of the
city and the neighborhood with a minimum of 30 respondents for
each neighborhood subtype. The Bureau of the Census designed the
sample for us in each city. Two questionnaires were then carried
out: a set of questions about the physical living conditions of the
family with an eye to detailed descriptions of house structure, housing materials, and house use and a second questionnaire that consisted of 233 questions (reduced to 213 in 1978) concerning the social and economic situation of the household along with a detailed
inventory of its membership and some attitude questions from its
principals. Heads of households or their spouses, but not children,
were asked to respond. We report on the data from the socioeconomic survey.

Neighborhood Types

Invasiones are those neighborhoods that are settled by an organized
incursion of squatters, or "parachutists" (*paracaidistas*), who set up
what are often called shantytowns, or "irregular settlements," and
attempt to squat on unoccupied areas in and near cities and hold
them mainly by moral force. Invasions are a well-known feature of
the Latin American urban real estate market and were one of the few
feasible housing strategies for the poor renters and immigrants who

either had come earlier to the city or were currently establishing themselves there, particularly during the great rural to urban movements of the 1950s and 1960s. Although the squatters were initially regarded with fear and distaste by the middle and upper classes, more enlightened analysts, such as John Turner (1968) and Anthony Leeds (1968, 1969), and politicians like President Luis Echeverría could see how much public money could be saved by having the poor house themselves at no cost to the state or their employers and, in their different ways, regarded the practice as a solution to housing in Latin America, rather than as a problem, an attitude that has trickled upward as the rural-to-urban movements slowed in the 1970s and 1980s. Marxist commentators could rightly grumble that the poor were being made to provide their own variable capital (i.e., subsidize the costs of their own reproduction and the reproduction of the labor force), but capital would come eventually to see that the practice was economically advantageous to them, even though, when they went home to their suburbs at night, they sometimes trembled at the prospect of the gardener acquiring a house lot for nothing and disturbing the bourgeois peace in the bargain. During the 1960s, and 1970s, periodic attempts were made to mobilize the squatters politically, more so in Colombia and Peru than in Mexico. More frequently the governing groups became uneasy and sent in troops to turf out the squatters. And often violence ensued.

The two most dramatic accounts of invasions in Mexico, which detail in different ways the desperation of the squatters, the boldness of their leaders, the constant danger and threat, and portray vividly the mixture of fear, anxiety, hope, pride, and desperation of the parachutists are Elena Poniatowska's chronicle of the invasion of the *colonia* Rubén Jaramillo (Poniatowska, 1980) and Carlos Vélez-Ibáñez's book on Ciudad Netzahualcoyotl (Vélez-Ibáñez, 1983). Poniatowska's is bloodier because the authorities of the state of Morelos felt enormous political pressure from Mexico City to expel the colonists, and the colonists were well organized and better led and armed than usual. The level of violence in political rhetoric and physical repression was very high: shoot-outs, police killings, trumped-up charges of drug trafficking against the colonists followed by the kidnapping of women and their consequent violation, the arming of neighbors, the seductions of government officials, constant harassment and intimidation, combined with the venality of the press, the ingenuousness of students, and the final crushing of the movement and killing of its leaders (or their imprisonment for 40 years) makes harrowing reading. The Vélez-Ibáñez account may be less sensational, but at the same time it gives closer attention to the "nuts and bolts of defending yourself" in a major invasion. It has heroic moments, as when the women of a neighborhood, acting on their own warning system, beat off an attack of government thugs and shady real estate dealers by stoning them.

Parachuting was never an ideal solution, since the potential for violence was always so high, and the colonists' ability to defend themselves was never absolutely guaranteed, and fortunately it is not so frequent in the late 1980s. We were somewhat relieved to find that, at the time of survey in Oaxaca, the area occupied by *invasiones* was about 1 percent of the total area of the city, and except for one group that took up temporary residence in the middle-class area of the city, parachutists were confined to areas in which house sites were of little value and would therefore not evoke the full repressive response of the authority. House and lot tenure in *invasiones* are, by definition, irregular. There is little permanent housing, many shacks, and all residents are extremely poor.

Colonias Populares

Colonia popular is the Mexican term for what is often translated as shantytown, a particularly misleading translation, since the term in English implies temporary structures and transient populations, whereas a *colonia popular* is inhabited by a stable population working on improving houses and lots whenever funds permit. Musgrove (1978) found that people invested windfall income in housing, and our interviews suggest the same. People will invest in housing when they can and when they must. They must invest in housing when they have none, or when they have an opportunity to get out of rental housing. But once they have a lot and a minimal shelter, they tend to lower their priority for housing dramatically and invest in it when they can. In Peru they were *barriadas*, and now *pueblos jóvenes*, while in Brazil they are still *favelas*; and in Argentina they are *villas miseria* or *villas de emergencia*, in a tone that mocks the government's attempts to define them as temporary. *Colonias populares* are permanent fixtures of Latin American cities; they may have originated in invasions, but now are stable communities and neighborhoods which have either obtained regular title from the government or are in the process of "regularization." More frequent are private housing developments on formerly undesirable house sites that have been sold for profit by their owners or agents. The houses in the *colonia popular* were "auto-constructed," and they resemble site-and-service projects in every way except for the absence of services in the first years, services that have to be wrung out of the city by political mobilizations. It usually takes about 10 years to complete the wringing process. Colonias are political units, and people in a *colonia* identify with it and themselves as members of it. The *tequio*, or Sunday communal work party (also known as the *faena*), is recruited from members of the *colonia*, and the *colonia* usually has a political apparatus (a *mesa directiva*) and elected officers.

One of the topics that we neglected to study was the urban real estate market. Our interviews do not show that people move about a great deal once they have located minimally decent housing in the city. Householders report that there is not any market in real estate, in the sense of a recognized equilibrium price for housing. A person who wishes to sell sets a price, and a buyer will either pay it or not. But very few wish to sell, if only because alternative housing for poor people is very difficult to find. We found that the peso values that people assigned to similar houses and lots varied substantially within the same *colonia* and that people showed very little propensity to move.[8]

We estimate that in 1980 *colonias populares* covered about 50 percent of the urban area in the cities that we were studying and that a majority of the population lived in them.[9] We divided the *colonias populares* into three subtypes: very poor, poor, and moderate-income neighborhoods. Very poor *colonias* were those in which the interviews showed households to have incomes at or below a single minimum salary.[10] The quality of life in these different popular neighborhoods is determined by age: the age of the householders and the age of the neighborhood. As we shall see in later chapters, there are very strong age effects in individual income profiles so that incomes peak around age 45, and there are very strong age effects in household income as well, since households with older household heads are much more likely to have older children who are active in the work force and therefore able to contribute to the household's budget. The higher-income neighborhoods tend to be older and to house older householders than the poorer. Similarly, older neighborhoods have had more time to organize themselves to obtain such urban services as roads, bus routes, running water, electricity, regularity of tenure, and perhaps even a clinic or elementary school.

Colonias populares have a life history, and after a while one can guess how long a *colonia popular* has been on its present site by looking at the kinds of houses that have been built and the kinds of services that they have. A young one consists mainly of bamboo shacks (*jacalitos*) and has few services. One that is 10 years old has between 30 and 50 percent adobe, brick, or concrete houses and also drinking water and electricity, the two highest-priority services. It usually has a politically active and aware citizenry, a good deal of internal political coherence, and a high degree of mobilization. By the age of 20, the houses will be mainly substantial multiroomed structures of adobe, brick, or concrete. The *colonia* will have most of the urban services that are available in settlements like it, but its level of political mobilization will be low, as will the effectiveness of its political structure. By the age of 20, the official party will have established itself and made itself the sole conduit of government grants and subsidies at the same time that it will have instituted party block organizations and assured the accessibility of the citizens to a police force that is beyond local control. The local organizations do not simply wither away: they are actively pushed aside, co-opted, and when necessary, threats and violence are used to ensure that the political organization that was necessary to acquire minimal urban improvements does not remain in place so as to form an alternative to the official party and to press for advantages (like equal access to education, or some form of political autonomy) which are "no business" of the local councils.

Stepick and Murphy (1977) have estimated that the difference between a site-and-service project and a *colonia popular* derives from the head start that a site-and-service project has in attaining the minimally decent standard because of the services it has from the outset. It can halve the time to become a minimally decent place to live.

Site-and-Service Projects

These projects are built by the process of *"auto-construcción"* (do-it-yourself home building). Sites are sold by a public agency that also sells services and building materials to the inhabitants. In Oaxaca people could buy materials and services if they wanted to, but the majority chose not to at the outset and instead built themselves a house that would not look out of place in a very poor *colonia popular*. A shack with earthen floor, makeshift windows, and a door that could in theory be locked, but in practice not, was not unusual. This was an unusual site-and-service project in the sense that most state directors would not have the commitment to democracy that the Oaxaca state director had. Normally directors would insist that some version of an architect's model of permitted housing be built, using official materials, even though the *colonos* would prefer to build to their own taste and timetable. Normally, people would be obliged to pay for and hook up to sewer connections as well.

If Mexican governmental institutions could handle money, which they cannot, site-and-service projects could well have been the solution to the Mexican housing problem during the years of the great migratory waves to the cities (1960–1980). But apparently no Mexican state or parastatal organization can manage the large sums of money and opportunities for graft that are involved in the sale and purchase of blocks of real estate. Such organizations and institutions are simply too vulnerable to outside political attack (and pilferage). Too many people have to be paid off. Directors who try to retain the funds for housing have been known to be threatened (unofficially) with official violence, so much so that they have had to fear for their lives and the lives of their families. People on the inside are also not without peccadillos. It is often difficult to convince a political appointee that money is for housing and not for him, for he has little interest in housing and enormous interest in feathering his own nest during the brief time allotted to him to do so under the Mexican system of sexennial replacement of officials and opportunistic peculation.

For these reasons, so expectable in the Mexico of the periods of Echeverría and López Portillo, the national effort at minimal aid for housing the poor failed to accomplish its goals.

Pueblos Conurbados

Pueblos conurbados are neighborhoods that were originally independent villages which were (mostly) involuntarily incorporated into the city. They are older, and many householders have quite substantial houses which they have been building for generations. Usually *pueblos conurbados* have a different feeling to them: former villagers feel more social solidarity and more community cohesion than the inhabitants of a typical *colonia popular*. They often have a long history of antagonism toward the municipality into which they were recently and unwillingly incorporated, and sometimes they have been able to retain the vestiges of independence from the city.

As part of their village heritage, they are usually much better organized than a *colonia popular* of equal age. Their Sunday *tequios* are better attended, although the concluding drinking fiestas are often better developed as well. There is a greater feeling of solidarity and neighborhood than in the average *colonia popular*. It is here that we are much more likely to find the networks of security of kinfolk and neighbor, and here as well "neighbor" can be a term with the force of "kinsman." They sometimes retain the residential and barrio structures of their village days, although village endogamy has disappeared, as has the practice of postnuptial virilocality that is still to be found in the villages. Still, one does tend to live near and with kinfolk, and there is still the village feeling that the people in the city are different, dangerous, and to be taken advantage of. Sometimes they are *colonias* of the city in name only and retain the agricultural, commercial, and artizanal base that they have had for years. But in this case change is on the horizon. The children, for example, will not work in the fields if they can possibly avoid it. They regard milpa agriculture carried out with the *yunta* technology[11] as a very distasteful way of making a living, and their parents, generally agreeing with them, do not require them to practice it. These more peripheral villages are being converted into dormitories for the city. In rare cases people are moving in and buying up houses and commuting to work in a car, treating the village as a place to have a weekend house (as in contemporary Tepoztlán, for which see Lomnitz-Adler, 1982). More frequently the villagers themselves have continued their process of semiproletarianization and begun to work in the city, retaining their interest and rights in village lands, but eking out their sparse *minifundista* incomes with earnings from exceptionally low wages or from profits from commercial activities.

Pueblos conurbados are not an especially Mexican phenomenon. They have been studied in North America and Europe, and their existence was important for understanding the structure of urban society and economy in Lima, Peru, as reported in Uzzell (1972) in particular.

The Central Area (*Centro*)

The *centro*, the oldest area of the city, is the place where most renters are usually found. Many recruits to invasions are householders who want to avoid paying rent, which, as we show in chapter 8, is a crucial saving in the household budget and without which saving, much opportunity for economic improvement is lost. If modest houses could be built and paid for in the central area, it would be an ideal place to live. All urban services are available there, and transportation costs for workers are reduced. Even with the undesirable rent burden, the center of the city is regarded as a desirable place to live. In Oaxaca the *centro* takes up about 20 percent of the city's area.

In Mexico City the *centro* is the most desirable place to live because it is subject to rent control, and people live there particularly in the *vecindades*, or apartment blocks made up of tiny one- or two-room units, almost free of charge. One of us made an anthropological pilgrimage to the home of Jesús Sánchez, the protagonist of *The*

Children of Sánchez, and found that the eponym had acquired neighbors undreamed of in that famous biographical document. The apartment houses were all gaily painted to conform to the sexual orientations of some of the more recent arrivals, and people who lived there reported great satisfaction with the place, particularly with the very low rent, the fabulous location, and the security that could be obtained in the world's largest city by locking the only door to the street. The difference between the scene depicted for the 1950s by Lewis and the one seen in the 1980s could hardly have been greater.[12]

Vecindades are a form of housing that has attracted much attention from social scientists, although not much has been written about them. Lewis's descriptions in the *Children of Sánchez* remain the best. They are usually very old, often huge colonial structures that have been converted in the years since the Revolution to tiny apartments. The apartments can be as small as one room, and the whole *vecindad* may consist of as few as 12 of them. The building is U-shaped, and there are six apartments on a side, with a common patio. These apartments are often called "*cuartos redondos,*" an expression whose original translation was "rooms with all conveniences." The patios, shared among all the families, are on most dry days filled with laundry on lines under which scores of young children are playing. The people in them tend either to be young and recent arrivals in the city or old and very poor, although we do not have direct statistics on their condition. The rent is very cheap. Even if the rent is not controlled, it runs to around US$15 a month. And given that they are usually centrally located, there are additional savings to be found in reduced transport costs, a major expense in all the larger cities. One of the most difficult problems for long-term residents in the *vecindades* is "*convivencia,*" or getting along with the neighbors. An unpublished study carried out by students in social anthropology at the Universidad Autónoma Metropolitana at Azcapotzalco compared the levels of consciousness about women's roles and activities as well as the degree of isolation of families in *vecindades* and in a *colonia popular,* hypothesizing that there would be much greater solidarity in the *vecindad,* where people lived on top of each other, than in the *colonia,* where the first thing one built (in the metropolitan zone of Mexico City at any rate) was a great *barda,* or wall, around the house which gave one security not only from thieves and burglars, but also from the prying eyes of the neighbors.[13] Not so. Even though the authors did not claim definitiveness for their study, their rough cut did not show any higher levels of *concientización,* or of solidarity, than did their colleagues in the *colonia popular.* Later when we lived in a two-room apartment in an eight-unit apartment house in Ciudad Netzahualcoyotl, we understood why. The problems of getting along with the neighbors are formidable when there are large numbers of children in small spaces. Kids naturally appropriate all the public space in the building, just as they would in the village, and equally naturally become embroiled with each other; and the families where we lived had great difficulty in settling on arrangements whereby one adult could be responsible for settling the inevitable scuffles and rows that occurred among the children. The exceptions to this rule occurred when all or many of the people who lived in the *vecindad* were related, which is not all that frequent, anymore at least, but even then we discovered instances of siblings who were not on speaking terms because of fights over the discipline of the children.

Middle-Sector Housing

Middle-class housing comes in two varieties: subsidized government-built houses (by state enterprises like INFONAVIT, which specializes in downscale middle-sector housing; ISSTE, whose houses are built for state workers and are usually much more spacious and better built; and INDECO itself, which built middle- and upper-scale housing in new neighborhoods), and owner-occupied houses built by the owners or by housing contractors in the fashion usual in the United States.

Both ISSTE and INFONAVIT build housing projects, and they are assigned to people by lottery, with fixed payment schedules. When inflation runs between 50 and 150 percent per annum, the value of these payments declines accordingly, and so being assigned a house or apartment is a considerable economic windfall. The windfall is sometimes compounded by the illegal practice of renting the newly acquired housing and staying in the house one has built in the *colonia*. But this is risky because anyone found out by the authority loses the right to public housing.

Public housing is not without its flaws. Finsten and Murphy (1988) interviewed about 40 householders in state-subsidized housing in Oaxaca and found that the families who were living in the projects were busily setting the interiors to rights. The architects had planned scaled-down versions of middle-class housing with separate bedrooms and special function rooms which used up a great deal of interior space and did not fit the needs of the families for common space in which to spread out. The authors comment (1988 : 4), "The desire for greater 'public space' is not particularly surprising. Most lower middle-class houses place greater emphasis on communal living area, often in the form of large rooms that function simultaneously as kitchens, dining rooms, living rooms, and sometimes bedrooms as well. In Oaxaca numerous bedrooms are a luxury of the upper and middle classes. When resources for housing are limited, people's priorities do not emphasize many single-use rooms. Instead they want space to mingle as a family, watch television and entertain . . . their family and friends." As a result, householders were busy pulling down partitions and making additions so that they could have a larger kitchen and living space, more on the model of the houses that they would build (and had built) for themselves in the *colonia.* All the subsidized houses have indoor plumbing, running water, and indoor bathing and toilet facilities, all of which are appreciated by the owners.

The middle-class housing projects are built for speculative sale and do not differ much from middle-class houses anywhere in Mexico. They too have their *barda.* Normally they have cramped quarters for live-in servants, even though the practice of live-in servants is slowly passing into desuetude. And even now once a servant girl becomes pregnant or marries, she is expected to form her own household and find her own place to live. These houses are usually smaller than U.S. houses, with large public rooms and small bedrooms. Although middle-class projects are better supplied with urban services, there is some democracy in Mexico. The electricity fails almost as often as in the *colonia popular,* and so would the water, except that the richer people have larger cisterns and more than one pipe leading from the water main into it.

These then were the neighborhood types that we studied in Oaxaca and, with local variation, all over Mexico.

The Survey Questionnaire

The questionnaire, consisting of 233 items in the Oaxaca version, was later modified and reduced to 213 questions. The longer form differed from the shorter one in that it had space for the inclusion of data on 12 household members, while the new form (adopted in 1978) had room for only 10. In the new form, as well, data on the household migrants was included. Both forms had four parts. Part 1 dealt with the members of the household and their characteristics (age, marital status, sex, education, ethnicity, place of birth, occupation, stability of work, and wages). Part 2 concerned household budget: expenses in 11 categories. Part 3 concerned the house, house lot, and neighborhood: how site and house were bought or paid for, what kind of materials went into the house, what plans existed for the im-

provement of the house, and what kind of amenities the neighborhood had. Part 3 dealt with householder attitudes toward their situation, the city, and the desirability of their housing situation compared
with the rest of the city as well as their neighbors. Questions were
asked about the householders' priorities, and we asked them to order
in importance such amenities as running water, electricity, security
of tenure, and the like as they saw each of these raising their standard of living.

The interviewers in Oaxaca were second-year students in the local
School of Social Work and were supervised by Murphy and Ruiz. In
the other nine cities, students were also used, as were local sub-
contractors, supervised by Ruiz, with assistance from Castañeda,
who, herself, supervised the design and data collection effort in Reynosa, Tamaulipas. The Oaxaca survey was designed in 1976 and carried out for the most part in 1977. It was followed by Villahermosa,
Tabasco; Reynosa; Mérida, Yucatán; and Venustiano Carranza, D.F.
(which is a *delegación* of the Federal District) in 1978. Finally in
1979, utilizing the shorter data collection form, Mexicali, Baja California; Querétaro, Querétaro; San Luis Potosí, San Luis Potosí; and
Tampico, Tamaulipas, were covered.

Field Work

Field work was carried out in five sites to complement the survey
data. Castañeda carried out field work in Reynosa to verify that the
analytic categories and typological methods worked there as well as
in Oaxaca. Cabrera did a historical and political study on Hermosillo,
Sonora, which although not part of the current study sample added
insight. He wrote his M.A. thesis on that city, where he now serves as
director of the successor institute of INDECO, which was called Secretaría de Desarrollo Urbano y Ecología (SEDUE) in the sexennium
from 1982 to 1988. Murphy has continued to do field research in
Oaxaca and has recently (1987) interviewed a sample of 660 households, 10 years later. Ruiz is carrying out field work on labor markets. Lorenzen and Selby with the assistance of Myung-Hye Kim and
Liliana Valenzuela, all of the University of Texas at Austin, did field
work in Ciudad Netzahualcoyotl, focusing on household management and small businesses.

Conclusion

When we began the continuing study, Mexico was preparing itself
for the administration of prosperity in the fatal phrasing of its presi-

dent, José López Portillo. The decennial follow-up was carried out in the fifth year of an economic crisis unprecedented in Mexico's history, with the possible exception of the period of the decade 1910–1920, the period of the Mexican Revolution. We began with the idea that we would provide, at the national level, a computerized data base with a full set of socioeconomic variables for planning and constructing the Mexico of the 1980s on a rational basis, with full attention to be paid to the desires, utilities, and conceptions of the ordinary families of Mexico. Bottom-up planning with a vengeance, as would suit the predilections of social anthropologists. In 1989 the president of Mexico promised to arrest the decline in incomes of the popular classes and promised as well not to make them suffer more for the calamitous policies that created the immense foreign debt that weighs so heavily on them. If the mood of the book is somber, it reflects the reality of life for ordinary people in Mexico's cities. If it seems to celebrate the "common man,"[14] then that is its intention, even though the sociological monograph is hardly the proper genre. We obviously admire the ordinary people of Mexico. But at the same time, we try not to be sentimental about them because they are a pretty hard-nosed bunch and very quick to take offense at the pretense of prettifying themselves, their families, their communities, and their lives. They also love to deflate the people who study and comment on their lives, like the citizen of Ciudad Netzahualcoyotl who was being interviewed by the *New York Times* reporter after the September 1985 earthquake and was asked:

NYT: Was there extensive damage then, in Cd. Netzahualcoyotl?
Citizen: No, we seem to have escaped the worst of the earthquake.

NYT: But you personally, you said that your house suffered some damage, did you not?
Citizen: Why yes, as a matter of fact, the roof fell in, while I was standing in the living room and it struck me.

NYT: Were you injured? Was your family injured?
Citizen: No! It was a cardboard roof.

2. Cities of the Study

D aily life in Mexico used to be very different in the different regions, just as speech patterns, culinary styles, and regional
characteristics are today. But however many differences appear to
exist among the regions as, for example, between the Americanized
regiomontanos of Monterrey; the conservative, traditional, warm
tapatíos of Guadalajara; the hustling, devious *chilangos* of Mexico
City; and the brusque, direct, and capable *norteños* of Sonora, the
fact of the matter is that for most people life in the cities of Mexico
is mostly the same. For the middle and upper classes, life may change
from city to city, but not for the majority of Mexicans. Obviously
climate mandates changes, and there is truly a great difference in the
quality, pace, and aggravation of living among the 18 millions of
the metropolitan zone of the capital as compared to the cities of the
provinces, as those of us who have had the dubious pleasure of living
in the poorer areas of that great urban space will attest. But even the
contrast between the capital and the provinces is easy to exaggerate.
It is striking how similar life in the cities of Mexico is for the poor
urban majority, involving as it does the same struggles and strategies, similar hopes and similar troubles everywhere.

But some cities are better than others for ordinary Mexicans, and
we take the opportunity in this chapter of introducing the reader to
the cities of the study by discussing the differences, particularly in
the way that these differences affect ordinary people. The key is the
job market. Cities differ (though not greatly) in the degree to which
they show potential for expansion in the number of good jobs in dynamic sectors of the economy, based on their past performance and
characteristics. We rank order the cities on a measure of labor market potential and categorize them into cities with comparatively
good or dynamic and active labor markets and cities with comparatively poor ones.[1] Then, we shall discuss the living conditions in
these two kinds of cities, finding, as might be expected, that cities

with good labor markets are nicer to live in than cities with poor ones. Last, we shall construct a "livability" scale, enabling us to rank the cities on living conditions as they affect ordinary people and to compare the scale with some others that have been constructed. Some preliminary cautions are appropriate so as not to mislead the reader. We shall not be discussing many of the features that are commonly included. For the majority of Mexicans, shopping conditions in the central district are irrelevant to their needs, since they have their own shopping places in the central market or in their neighborhood. The availability of entertainment aside from movies is irrelevant as well, since ordinary Mexicans do not patronize restaurants, clubs, concerts, or sporting events anymore than they take advantage of the availability of CAT scanners or high-speed computers.[2] Entertainment is provided by one's self and one's family, or it is confined to television and the occasional wrestling match. Family fiestas celebrate the best in ordinary people's lives, particularly when they are supplemented, in good times, by drinking among *cuates* and visits among neighbors and relatives. What matters to ordinary Mexicans is home and job, life and its sustenance, and in this the Mexican urban experience tends toward homogeneity.

The Urbanizing of Mexico

Mexico is now an urban country since over 50 percent of its population lives in cities. By 1970, 53 percent of the population lived in communities of 5,000 or more, and 45 percent lived in communities of 15,000 or more. By the time of the INDECO study, there were 73 cities in Mexico with populations of 50,000 or more. Mexico has turned into an urban nation very quickly. At the time of the Revolution in 1910, Mexico was a predominantly rural country where only 11 percent of its population lived in communities of more than 15,000 inhabitants, and Mexico City had a population of three-quarters of a million, dominating all other cities by far. The rate of urbanization increased steadily during the period since the Revolution: in the 1920s it was 3.5 percent per annum; in the 1930s, 3 percent; in the 1940s it jumped to 5.9 percent; in the 1950s, 5.5 percent; and in the 1960s, 5.4 percent. These growth rates can be compared to the growth rates of the general population in Table 1.

The causes of urban growth are to be found in the displacement of small landholders from their plots in the first five decades after the Revolution, which created the massive waves of migration that peaked in the 1960s. Since then natural increase has accounted for the majority of population growth in the cities. Today 70 percent

Table 1. *Population Growth Rates (In Percentages)*

Years	Growth Rate Total Population	Growth Rate Urban Population
1900–1910	1.1	2.2
1910–1920	−0.5	1.5
1921–1930	1.6	3.5
1930–1940	1.7	3.0
1940–1950	2.7	5.9
1950–1960	3.0	5.5
1960–1970	3.4	5.4

Source: Unikel, 1976:30, passim, and *Anuario Estadistico*, 1984, Table III, 1.3, pp. 141–146.

of the increase in the urban population is due to natural increase, and the most recent figures show a slow decline of the rate of rural-urban migration through the 1970s, climaxed by a decided downturn in the middle 1980s as the economic crisis reduced the rate at which urban jobs were being created, so much so that it was being reported in the press that some families had given up in despair and had returned to their home villages. Since the administration of Miguel Alemán (1940–1946), heavy investments have been made in agriculture in selected areas (notably the northwest and northeast), and as a result, in these areas commercial farming has prospered. Mechanization has advanced, changing crop types from peasant and wage foods to luxury, export, and industrial foods and crops. Much labor has been "released"; people have been driven out of commercial agriculture and into the cities where some employment was being generated by the 40-year economic expansion (1940–1982). This was particularly true in areas like Sinaloa and Tamaulipas and, more recently, Baja California. In areas where small-scale agriculture predominated, like the Bajío, Michoacán, Jalisco, and Oaxaca, population growth which had been useful in past eras because of the high labor demands of small-scale subsistence agriculture was no longer justified because the resource base of the traditional communities and small holders had been eroded. This was true even in those areas where land reform had had a significant allocative effect. In many parts of Mexico today, the more "backward parts" to be sure, there are literally thousands of villages that are unable to feed themselves through full-time agricultural production because of the lack of land, or of good land, credit, and adequate technologies.[3] These populations, which have been rendered redundant by development, have had to move to the cities for survival, and it is in the cities that we pick up their story in this book.

During the period of economic expansion in Mexico (1940–1982) when aggregate economic growth averaged 6 percent annually, the cities were, with difficulty, able to absorb part of the growing rural population. Of course, the term "absorb" conceals a multitude of evils, including unsanitary living conditions, lack of rudimentary amenities, constant economic and political threat, unemployment and social instability, and working conditions equaling the worst working conditions in nineteenth-century Europe. The immigrants and their city-born children had, as the Marxists say, to provide their own variable capital, that is, pay the costs of their maintenance and upkeep from their own efforts, since their wages were entirely inadequate to cover the costs of housing and subsistence. As we mentioned in the last chapter, what goes by the name of "invasions" or "parachuting" applies clinical nomenclature to what were often painful, rending, and anxious periods of people's lives as they were forced to take terrible risks for the meanest of livings. The experience of laying claim to a tiny lot, building a tiny shelter out of waste materials in furtherance of one's claim, and enduring the first perilous weeks of tenancy was an adventure to be enjoyed only well after the event. Violence could erupt at any point in the process. Most frequently the violence was not official so that one could have some official recourse, but semiofficial, that is, fomented and organized by the authorities but carried out by private instrumentalities, bullies, paramilitary, or real estate speculators' thugs, or, often indistinguishable, real estate speculators themselves. Violence could erupt among householders as plots were sold to many people at the same time, and claims could only be enforced through strength and self-help. And parachuting is an uncertain prospect at best, usually requiring years before clear title can be secured. And during the interim, the mud, the garbage, the darkness, the isolation, and the fatigue combined with the ever-present threat of dislocation, robbery, and violence exaggerated an already well developed suspiciousness on the part of poor householders.

It is hard to remember sometimes when one is studying the hardships of creating a niche in the city that life was worse in the *rancho*. Health conditions were worse than the most crowded city slum because of the prevalence of insecure and polluted water supplies.

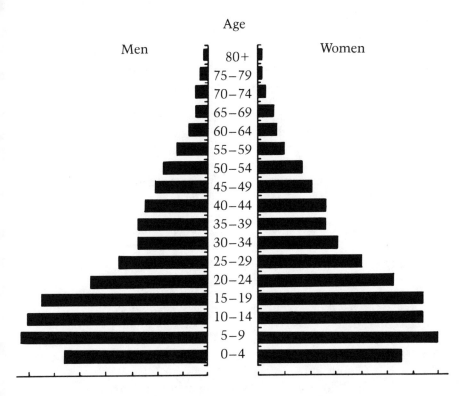

Figure 1   **Population Pyramid of 10 Mexican Cities**

Medical attention was generally not available and when available was beyond the resources of most. When food was short, malnutrition was common. In the villages of Oaxaca, at the end of a drought period in the 1960s, fully 30 percent of the villagers were reduced to a diet devoid of meat, beans, and coffee, which was described accurately by the villagers as one of tortillas and salt, with a little chile rubbed on the tortilla. Sickness was pandemic. In the cold season or in the highlands, bronchial infections predominated, while in the wet season the diarrheas were fatal. Children died of measles and dehydration in the dry season, and in the wet season dehydration, enteric conditions, and worms took terrible tolls. Fully 50 percent of the village children died before the age of five as late as in the 1970s in Oaxaca. For Oaxaca these conditions have been demonstrated by Graedon (1976), Kappel (1976), Selby (1974), Malina and Himes (1978), Malina et al. (1980a, 1980b).

Urban invasions are dramatic and eye-catching, but most poor people who now live in the cities did not invade the lots they now own, or even their first. They came peaceably to the cities and shared rooms or houses with relatives, then found their own place in the city, often first in rented quarters, either downtown or on the outskirts. When they could, they moved to the *colonias populares* that were opening up at the edges of the cities, purchasing their lot and defending it to the best of their ability. Legal *colonias* were founded to house urban workers, particularly during the 1960s, when the rate of population increase was the greatest. By now the urban birth rate has begun to decline so that urban growth has finally come to be somewhat restrained. Figure 1 shows the population pyramid for 10 cities of Mexico as drawn from our data.

The population is young, with a median age of 16.6 years with a slight deficit of women over the age of 65 and under the age of 15. The most significant changes in the population took place in the youngest birth cohort. When one compares the cohort born in 1972 to that born in 1977, the younger is 22 percent smaller than the cohort that preceded it, which was the largest cohort of all. This compares with a 9 percent drop in the population as a whole, according to the not terribly reliable census figures for 1980 (SPP, 1986:29). Except for the decline in the most recent cohort, the pyramid is typical for a developing country. The recent declines in rates of natural increase are important and signal the beginning of a slowdown in the rate of growth of the urban population.

Cities of the Sample

Nine cities and one *delegación* of Mexico City provide data for this study. Table 2 presents them in order of population size.[4] This is not a random sample of the cities of the republic, but the sample does provide reasonable coverage of the different types of urban situations. Two are from the northeast (Tampico, Reynosa), two from the northwest (Mazatlán, Mexicali), two from the frontier areas (Reynosa, Mexicali), two from the central region (Querétaro, San Luis Potosí), one from the southeast (Villahermosa), and one from the Yucatán (Mérida). The nation's capital is represented by the largely working-class delegation of Venustiano Carranza. Five of the cities are state capitals, as were half the cities over 50,000 at the time of the survey (Mérida, San Luis Potosí, Querétaro, Villahermosa, and

Table 2. *Population and Sample Size for*
Cities of Sample

City	Population	Sample Size
Mexico City	8,831,079	1,134
Mexicali	510,664	831
Mérida	414,529	569
San Luis Potosí	406,630	1,013
Querétaro	293,586	821
Tampico	267,586	821
Villahermosa	250,903	265
Mazatlán	249,988	924
Reynosa	211,412	1,688
Oaxaca	157,284	1,386
Total		9,458

Oaxaca). We feel that the variety is sufficient to enable us to discuss urban life in Mexico, at least in its general outlines, despite the inadequacies of the opportunistic sampling strategy.

Types of Cities

In one sense there are only two kinds of cities in Mexico: Mexico City and the others. This is reflected in the contrast that people in Mexico City make between "here" and *la provincia*, a term that ordinary people use with no pejorative connotation.[5] The role of the capital city, as primate city in the republic, and the difficulties associated with its enormity of scale as it surpasses the 18 million mark in population (counting the surrounding contiguous populations of the state of México into the metropolitan zone, as is usually done) is well known. Mexico City is the manufacturing center of the country. About 60 percent of the manufactured goods in the republic are produced there. Two of the largest cities in the republic are found within the metropolitan zone, the capital and Ciudad Netzahualcoyotl, which is estimated to have a population between three and four million people. The metropolitan zone is the consumption center of the country as well. All roads and rails lead to Mexico City, and attempts by several administrations to reduce the number of people in the area have been largely futile, although each administration does implement a decentralization plan. (Typically, however, in recent years, the national oil company, which was traditionally located in both Mexico City and Tampico, chose to build its huge office tower and headquarters in the capital's downtown area but out of deference to the decentralization plan which frowns on the circulation of even more vehicles in the federal capital, declined to build parking facilities, creating a traffic snarl of monumental proportions, even by Mexico City standards.) There are more than two million vehicles registered in the metropolitan area, and no one knows how many unregistered ones, and since they have aged a good deal during the economic crisis, their contributions to the already dreadful pollution levels have increased markedly in the past five years, 1982–1987. Mexico City is the most polluted city in the world. Mexico City, with its metropolitan zone, is also the most populous city in the world. It has been admirably described by a subsecretary of Housing and Human Settlements as a city which is *"chata, chaparra, y cacariza,"* meaning "flat, squat, and pock-marked," which description is even more accurate after the 1985 earthquake than it was before, when it was first rendered. It is an enormous expanse of small houses and low-rise buildings, stretching as far as the eye can see and pock-marked by abandoned building sites or sites under construction. Even before the earthquake there were parts of the metropolitan zone that resembled a bombed city more than the world's largest.

Aside from the monstrous primate city, Mexico has two other regional cities of importance: Guadalajara and Monterrey, and these vie with Mexico City for domination of their own hinterlands, as well as for a say in national development policy. Before the collapse of the Garza Sada empire in Monterrey, northerners contested national policy and fought hard for independence from the capital which they regarded as dominated by meddlesome bureaucrats and consumed with political fiddling, rather than straightforward, technically minded, and efficient executives on the perceived American model. The decline of Monterrey has been offset by the rise of Guadalajara, which has succeeded better than any other large city in maintaining economic progress during the current period of economic crisis. Like Monterrey, but now more so, Guadalajara has been able to achieve some degree of financial and political independence from Mexico City.

Regional Cities

Despite the high degree of urban primacy, Mexico does have its regions, and it is important to recognize that regional cities can be dominant in their own areas. Thus Mérida gains special status as the regional city of the Yucatán, as does Oaxaca in Oaxaca. Tampico–

Ciudad Madero is important in the northeast, as is Mexicali-Tijuana in the northwest, just to cite the cities in our survey. But one must be careful not to overestimate the role of regional cities. As Unikel's (1976) work on intercity exchange, and as economic analyses of the relative autonomy of Mexico's regions has repeatedly shown (Stern, 1973), the regions are not characterized by a high degree of political or economic autonomy. But still, Tampico–Ciudad Madero was the center of the oil industry from the beginning of this century until the present period, when production in the Bay of Campeche has superseded it. It is still the headquarters of the very powerful union and transnational corporation, the oil workers of Mexico (*los petroleros*), and the favored position of the oil workers in the Mexican economy for the past 40 years is mirrored in the much higher standards of living and levels of urbanization to be found in that city.[6] The famous "strip of gold" (*faja de oro*), the most significant oil find in Mexico during the early years of this century, yielded fortunes to the British, then to the American, and finally to the Mexican companies that exploited it. But as the symbolism of the Pemex tower in Mexico City shows, even without the well-known battles between the federal executive and the oil union, oil politics dominate production, and the domination is levied from Mexico City, denying Tampico the kind of regional quasi autonomy that it might otherwise have.

Mérida was an important regional political center in a region that fought the central government throughout the nineteenth century. It rose to economic prominence at the turn of the century, and particularly through the period of World War I, on the basis of henequen production in its hinterland. But with the decline of the importance of natural fibers and the increasing ability of the central government to exert control through the development of the centralized presidential system after the Revolution, its status as a semiautonomous regional center declined.

The fortunes of Oaxaca as a regional urban center have risen and declined in response to the degree to which its role has been usurped by Mexico City (see Murphy and Stepick, in preparation). During the early years of the century, it was a relatively populous, relatively autonomous center, isolated from the rest of Mexico because of the absence of road transport. The railways went through at the end of the nineteenth century, but Oaxaca was not adversely affected until the 1920s, when a very severe earthquake reduced its population considerably and began the downward trend in population and political and economic importance that continued until the 1970s. The arrival of the Panamerican Highway and the construction of air links to the rest of Mexico damaged Oaxaca, since capital flight joined the flight of people from the area, and as a result, from 1940 to 1960 (roughly), Oaxaca came to be dominated by national and international capital based in Mexico City. In the last two decades, there has been a reversal of this trend. Recently Oaxaca appears to be weathering the economic crisis rather better than the capital, and the government's decentralization program is finally beginning to send numbers of well-paid, well-educated middle-class bureaucrats to the city.

The regions of Mexico are defined culturally and environmentally more than they are economically. Unikel (1976) divides Mexico into eight socioeconomic regions: the Northeast, North, Gulf area, North Central, West Central, Central, Valley of Mexico, South, and Southeast, but he is unable to bring convincing data to bear in support of the notion that they are relatively autonomous. After some brief, descriptive remarks—indicating that, for example, the Northeast, North, and Valley of Mexico have gained population in the period 1900–1970, and West Central and Central have lost it; that the Valley of Mexico is the most highly urbanized, and the South-Southeast, the least; that the Gulf area is the region that shows the greatest proportion of cities that have gained population compared to those that have lost—the analysis loses coherence and founders. The regions are geographical outlines; the relations that characterize them are historical happenstance; and the whole analysis is not terribly revealing. His analysis of the national urban system is somewhat more satisfactory. He creates urban hierarchies within the regions that are familiar to the Mexicanist—such as the Bajío, Jalapa-Orizaba-Veracruz, Monterrey, Guadalajara, Tijuana-Mexicali—and defines three categories of cities: dominant cities (Mexico City), subdominant cities (Guadalajara, Monterrey), and dependent cities (the rest). He uses a gravity model based on intercity cargoes to operationalize his typology and, indeed, is able to show that the dominant city dominates, the subdominant cities do dominate their hinterlands and exchange with each other as well as the capital city, while the dependent cities exchange mostly with their hinterlands and the capital. Also Mexico City is linked by cargo to every city in Mexico, as are Monterrey and Guadalajara to a lesser degree, while the dependent cities show only regional connections.

Our Typology of Mexican Cities

Our approach to typing Mexican cities is not regional or even strictly economic. We try to capture the urban reality from the point of view of ordinary people who live in the city. Our interviewing in the cities had started as early as 1970, when we interviewed urban house-holders in Oaxaca about their reasons for coming to Oaxaca, or for wanting to move somewhere else, particularly in those days, to booming Mexico City (Murphy, 1973; Stepick, 1974). In that earlier work we had asked about the reasons people had in moving to the city and how they thought about the various cities of Mexico, asking questions like "What are the advantages and disadvantages of living here in the city compared to where you came from?" and "If you were able to pick a city to move to from this list, which one would you pick and why?" Ordinary people have an inevitably partial view of the national urban system, but they were able to compare life where they were with the potentialities in living in other cities. When the ordinary Mexican evaluated different cities, two kinds of criteria were used: labor market conditions, on the one hand, and the general living conditions, on the other. We tried to capitalize on their insights and to represent them in our measures of the overall desirability of the cities in our survey. Labor market conditions were inferred from data in the 1974 industrial census, while the "livability" of the city was determined from our own data.

Labor Market Conditions

The best measure that we could devise to evaluate the job creation potential of a city was one derived from the sectoral attributes of the work force.[7] Cities in which there was a relatively high proportion of workers in the most dynamic subsectors of the manufacturing sector were deemed to have more growth potential than those with lower percentages. More specifically, our labor market condition index was the product of the percentage of the labor force in the most dynamic and expanding subsectors of the manufacturing sector, as a proportion of the total employed work force in the city.

The most recent industrial census to which we had access was the Censo Industrial of 1974, since the 1981 Industrial Census was not available at the time of writing. We typologized our 10 cities according to two criteria: the percentage of the work force engaged in manufacturing and the percentage of the work force engaged in sectors classifiable as both "active" and "dynamic." A subsector was said to be "active" if the rate of growth of the work force in that subsector exceeded that for the sector as a whole, and a subsector was classified as "dynamic" if the rate of growth of output for a given subsector exceeded that of the sector as a whole. We used the definitions of "sector" and "subsector" provided by the Censo Industrial.

The following subsectors were classified as active and dynamic:

Intermediate goods
Capital goods
Consumer durables
Chemical products
Nonmetallic mineral products
Basic metal industries
Metal products
Manufacture and repair of electrical goods
Manufacture and repair of vehicles
Paper and paper products

Table 3 gives the scores. Partitioning the cities at the median value of 7.5 yields a classification in Table 4, albeit crude, of "high growth potential" versus "low growth potential" cities.

This partition coincides with our intuitive feelings about energetic, "bustling" cities compared to more stable, "sleepy" cities, although these descriptions should not be taken to imply that the more stable cities are incapable of population growth, for that would be a mistake. They have experienced population growth in the past two decades. Leading the cities with good growth potential, that is, with a relatively high proportion of manufacturing jobs and jobs in the most dynamic subsectors, is Mexico City with its highly developed industrial, commercial, and service sectors; it is both the administrative and the industrial nerve center of the country, with the

Table 3. *Index of Growth Potential in Labor Market*

	Work Force in Manufacturing %	*Work Force in Dynamic and Active Sectors %*	*Index*
Querétaro	29	62	0.18
San Luis Potosí	25	40	0.10
Mexico City	23	34	0.10
Tampico	14	61	0.09
Mexicali	23	34	0.08
Mazatlán	31	19	0.06
Mérida	22	18	0.04
Reynosa	13	27	0.04
Villahermosa	14	22	0.03
Oaxaca	20	16	0.03

Source: Censo Industrial, 1974.

Table 4. *Labor Market Potential for*
Sample Cities

High-Growth Potential	Low-Growth Potential
Querétaro	Mazatlán
San Luis Potosí	Mérida
Mexico City	Tampico
Tampico	Villahermosa
Mexicali	Oaxaca

fourteenth largest economy in the world in aggregate terms. Mexicali is an important transport center, an increasingly important site of twin plants, or *maquiladoras*, which have been the most dynamic centers of production and employment through the economic crisis that began in 1982. It is also the center of agricultural development of increasing importance for export agriculture (the world's greatest producer of asparagus, e.g.) and an important staging area for emigration to the United States. San Luis Potosí is a recent member of the group of dynamic cities and perhaps the most surprising one. It has always been an important transportation center and in recent years has become an important manufacturing one as well. A good deal was invested in the city and the region during the Echeverría sexennium (1970–1976), and although it is 450 miles from the northern frontier, it has been declared a "border area" for the purposes of twin-plant production. The rate of growth of the industrial zones to the east and south of the old city are impressive even to the casual eye. There has been much job creation immediately before and during the period of this study.

Querétaro has long been a manufacturing center and has grown, particularly since the construction in the 1960s, because of the expressway link with Mexico City. It has a long history as being the commercial and political center of Mexico's historically most productive agricultural region, the Bajío, as well as an important center of political activity before and during the struggle for independence. Many industries have found it efficient to put branch plants in Querétaro, rather than in Mexico City, since Querétaro is well situated to serve the vast industrial and consumer market of the metropolitan zone of the capital. The final city with high growth potential is Tampico, the city of the *petroleros*, the oil workers union turned transnational corporation; it has enjoyed the benefits of oil development since the late 1930s. The oil workers and their officers have long been an economically favored group of Mexican labor, and the city has developed into a reasonable place to live for those who can get jobs with the union or the company. Since the union tends to favor nepotistic recruitment, those families in Tampico that have a connection with the oil union have tended to be more stable, and their members have been more inclined to invest in their city and their homes.[8]

Cities with Low Growth Potential

The cities with relatively smaller proportions of the work force in dynamic and active subsectors of the manufacturing sector are much as one would expect, with the possible exception of Villahermosa, which has become the administrative center of the oil development in the Bay of Campeche. During the period of the study, however, Villahermosa had not yet benefited from oil development, rather it had been assaulted by it. Its infrastructure was being stretched to the limit or ruined as huge Pemex trucks attacked its roads and Pemex personnel and business activity overwhelmed its communications. Oil exploration with little care for economic or ecological damage was ruining not only fisheries but also agriculture. For Tabasco, during the period of the study, petroleum development was a disaster (see Hellman, 1983). Local inflation was very high, and Pemex was importing workers either in response to union demands or in response to traditional administrative practices as dictated by Mexico City. The amount of money that was being put into the local economy by Pemex during this period was very small indeed. Oil development represented a net loss for the local area. (This changed thanks partly to the angry and effective intervention of the governor of the state of Tabasco.) The other surprising member of the low-growth category is the city of Mazatlán, and it might be remarked that it is only a marginal member of the category.

Livability of the Cities

In our interviews in the five cities where participant observation was carried out, ordinary people emphasized three themes which, for them, defined the desirability or "livability" of the city in question: income and household budgets, housing and urban services, and, as expected, jobs and employment conditions. In addition, we had survey items that asked people why they had come to the city. These latter data confirmed the interview material, in that, for example, 65 percent of the householders interviewed placed jobs at the top of their list.[9] Family reasons are given by 12 percent of the people in the survey who had moved to the city in which they were now living. This meant two things: the preservation of the family, keeping it together, as a positive goal in itself and utilizing family members as a resource for finding work in the city and for temporary lodging upon one's arrival. Education was also given as important to the householders, though only 6 percent listed it as first in their minds, as was home ownership, with 6 percent listing it as first in importance.

We also asked people about their reasons for moving within the city itself, which 48 percent of the households had done since they had established themselves in the city in which they were currently living. The most important reason for moving around in the city was to secure a homesite: 56 percent of the householders reported home ownership as the most important reason for moving within the city. Jobs were given the highest importance by 20 percent of the householders, with three-quarters saying that they had moved in order to take advantage of a better paying job opportunity, while one-quarter suggested that it was the availability of work that was important for them. Family reasons were given by 6 percent of the householders; the availability of urban services was given by 4 percent; and educational opportunity was given highest importance in determining a move by next to no one (0%).

The themes that emerged in interview and survey were the expected ones: work, jobs and wages, family preservation, education, home ownership, and the availability of urban services. It is on the basis of these topics that we selected variable sets to define our scale of "livability" as the second dimension of our study of urban life. When comparing cities with poor and good labor markets, one can see that on average better labor markets go hand in hand with more livable cities.

Income and Household Budgets

Household heads earn more and households as a whole have higher incomes in the high-growth cities (see Table 5 for data on income and household budgets). When we compare the median incomes of heads of households, we notice that high-growth cities show an increase of 26 percent over those of low-growth cities, while household incomes are 19 percent higher.[10] Per capita incomes are 13 percent higher despite the fact that households are larger in the high-growth cities, with 5.5 members as compared to 5.2 members in the lower-growth cities.

Table 5. *Livability: Income and Household Budgets*

Variables	Labor Market Potential of City	
	High	Low
Median income household head	$3,790	3,000
Total household income	4,079	3,419
Per capita income	741	657
Food expenditures (monthly)	1,650	1,947

Note: Figures are in 1978 pesos.

The main budgetary item for the ordinary household in urban Mexico is food, and the high-growth cities show that, although households spend 16 percent more on it, the expenditure is a slightly smaller proportion of the household budget.

Housing and Urban Services

Housing conditions are important to ordinary Mexicans. Many have had to fight hard to obtain minimal housing, and they have had to defend it courageously against the outside. Being able to secure housing is of paramount importance. Next in importance is securing regular title to one's house and lot. After that comes securing such basic urban services as drinking water, electricity, and sewer connections. In all of these areas, people in the high-growth cities have been able to do better than people in the low-growth cities, as Table 6 indicates.

The stability of households in the city (in the sense of not having to move house) is important to ordinary people, and households in the cities of high growth potential do better than the low-growth ones. More of their households were formed in the city (69% vs. 62%, e.g.), and more of the households have lived in the house they were interviewed in (55% vs. 50%). These advantages, small as they are, cannot be ascribed to longer residence in the different city types, for they are indistinguishable in this regard: around 60 percent of the households have been constituted in the last 10 years; 32 percent, in the last five years; and around 7 percent, in the last year. But still, their tenure is more secure; 81 percent of the households in the high-growth cities have regular (i.e., legal and enforceable) title to their houses and sites whether owned or rented, compared to 69 percent in the low-growth cities.

Table 6. *Tenure and Urban Services*

Variable	Low Growth %	High Growth %
House and site		
Households formed in city	62	69
Households with regular title		
(owned or rented)	69	81
Urban services		
Drinking water on housesite	71	91
Electricity	82	94
Sewer connections available	43	82

The cities with high labor market growth potential have extended urban services to more householders than in the low-growth cities. Drinking water, electricity, and sewer connections are all more frequent, although, in the case of the last, even in the high-growth cities, it only reaches the 82 percent level (as compared to 43 percent in the low-growth cities).

Housing amenities are more frequent in the households of the high-growth cities as well. Kitchens and bathrooms are more frequent, as is the availability of drinking water, with 90 percent or more of the households in the high-growth cities reporting all three, and only running water being in shorter supply in the low-growth cities, where 71 percent of the households report its availability.

Jobs and Employment Conditions

Open unemployment is not widespread in Mexico; it varies between 4 and 6 percent, indicating that Mexico formally, at least, has full employment. And it is true that everyone works, and everyone earns money, but this is more a necessity for survival in the absence of social security schemes than it is the availability of good jobs. Mexico's problem, like that of most developing countries is underemployment, or "disguised unemployment." This can be calculated in two ways: full-time workers who earn under the minimum wage or full-time workers who do not have registered formal sector jobs. In both cases the cities' job profiles show deficits, and in both cases the high-growth cities are better served, as one would expect, and as the data presented in Table 7 indicates. In high-growth cities 59 percent

Table 7. *Jobs and Employment Conditions*

Variables	Low-Growth Cities %	High-Growth Cities %
Job conditions		
Household head has registered job	51	79
Household head earns less than minimum salary	29	22
Household head has employment stability	63	81
Household head has fringe benefits	69	81
Job types of head of household		
Agricultural	5	4
Casual labor	11	5
Low-level service	14	22
Blue-collar	21	18
White-collar	40	40
Professional	10	11

of the household heads have registered formal sector employment, and 78 percent earn more than the minimum salary, while in the low-growth cities 51 percent of the household heads have formal sector registered jobs, and 71 percent earn more than the minimum salary.

Stability of employment is also important, as is the availability of fringe benefits, which supposedly are attached to all formal sector employment. In both job stability and fringe benefits, the jobs in the high-growth cities fare better than the jobs in the low-growth cities: 81 percent of the household heads in the former have achieved job stability in the sense that they have held their present jobs for at least one year, while only 63 percent have done so in the low-growth cities. Fringe benefits are enjoyed by 69 percent of the household heads in the high-growth cities and by only 51 percent in the low-growth cities.

The frequency of broad categories of employment in high-growth cities does not differ a good deal from low-growth cities. High-growth cities have more service and professional jobs and fewer agricultural, casual, and blue-collar jobs, but white-collar jobs exist in the same proportion in both types of city. The differences in the distribution of types of jobs are small.

The "Livability Scale"

It is possible to rank the cities within each category (and overall) on their livability on the basis of a "livability scale." We selected eight variables to describe the "livability" of the cities and present the data in Table 8.

1. Wages. Cities were evaluated on the percentage of household heads who had wages below the minimum salary for that region.
2. Number of labor force participants per household.
3. The value of the house site and improvements.
4. The percentage of households in the city who enjoyed legal tenure for both house and lot.
5. The percentage of houses with sewer connections in the city.
6. The level of educational attainment in the household, for both adults and children.
7. The percentage of matrifocal households in the city.
8. The percentage of complex households. These are households which are either lineally extended (half the cases) or which incorporate a distant relative or nonrelative (other half). In both cases they are efficient wage-earning and expense-sharing collectives and are for that reason desirable alternatives.

Discriminant Function

In order to assess the relative contribution of each variable to the dimension of "livability" and to find out how the cities scored on the "livability index," a one-dimensional discriminant analysis was done. This produced discriminant function coefficients which weight the variables as follows (from high to low), as is seen in Table 9.

Urban infrastructure and low wages carry the most weight in defining "livability," and the only counterintuitive ordering to be noted is that of "regular title" which like subpar wages has a negative value. Why this is so can be deduced by inspecting Table 8, where it will be seen that the two cities with the lowest rates of regular title and tenure (Mazatlán and Tampico) are otherwise distinguished by very high "livability" as is shown in the following city values where both cities prove comparatively livable despite low frequencies of regularity of tenure.

The canonical discriminant function values, evaluated at the mean, yield a rank order of "livability" of the cities in our sample. These are given in Table 10. The results can be discussed briefly. First, it is well to acknowledge that they are very similar to results

Table 8. Eight Variables Used in Livability Scale

City	Low Wages (Head of House) %	Number in Labor Force	House Value (Mean) $	Regularity of Tenure %	Sewer Connections %	Educational Attainment (Average Years)	Matrifocal Households %	Complex Households %
Mexico City	11	1.5	6,483	89	96	9	7	19
Mexicali	8	1.2	14,002	61	67	8	9	12
Mérida	21	1.4	9,699	77	20	7	7	24
San Luis Potosí	6	1.5	13,426	84	94	10	9	30
Querétaro	11	1.4	9,676	61	71	9	6	9
Tampico	8	1.4	3,850	54	71	8	8	15
Villahermosa	35	1.5	6,611	73	70	8	11	26
Mazatlán	5	1.3	15,898	52	75	9	4	17
Reynosa	38	1.3	5,498	65	35	5	5	10
Oaxaca	37	1.4	10,725	66	26	7	8	17

Table 9. *Canonical Discriminant Functions*

Variable	Function Value
Sewer connections	0.85
Value of house and lot	0.11
Complex household	0.10
Educational attainment	0.06
Matrifocal household	0.06
Number earners	0.01
Regularity of title	−0.09
Low wages	−0.39

Table 10. *Function Values for Cities in Sample*

City	Value
San Luis Potosí	0.90
Mexico City	0.76
Mazatlán	0.54
Tampico	0.33
Querétaro	0.32
Mexicali	0.32
Villahermosa	0.03
Mérida	−0.81
Reynosa	−0.81
Oaxaca	−0.87

of similar kinds of tests carried out first by Unikel (1976) and subsequently by Conroy et al. (1980). Unikel (1976:108−110) created a *"nivel de vida"* scale based on the level of education of the city dwellers, the percentages with piped water, sewer connections, bath and modern kitchen, and the percentage in any given city that wore shoes. All data were taken from the 1960 census. Conroy's (1980) "index of socioeconomic opportunity" was based mainly on data from the 1970 census and consisted of 10 variables: seven economic and three social. Economic indicators were (1) the proportion of population engaged in agriculture, (2) the agricultural output per employee, (3) the proportion of the labor force in industry, (4) industrial production per employee, (5) income per capita, (6) real wage for

Table 11. *Sample Cities on Three Scales*

City	Our Livability Scale	Unikel's Nivel de Vida Scale	Conroy et al.'s Opportunity Scale
San Luis Potosí	1	4	10
Mexico City	2	1	1
Mazatlán	3	5	6
Tampico	4	2	4
Querétaro	5	9	5
Mexicali	6	7	2
Villahermosa	7	8	8
Mérida	8	3	3
Reynosa	9	8	7
Oaxaca	10	6	9

low skilled, and (7) the percentage of the labor force employed. The social indicators were (8) percentage of houses with piped water, (9) number of physicians per 10,000 population, and (10) number of hospital beds per 10,000 population. Table 11 compares the rank orderings of the three scales on our sample cities.

In the 30 years that passed since Unikel's data was collected and the 20 years since Conroy's, real changes have taken place in some of the cities of Mexico, and none quite so evidently as in San Luis Potosí. At the beginning of the 1960s, it was in decline. The population growth rate for 1950–1960 was four points below the national average of 35 percent, but then in the next two decades, it soared to 47 percent (1960–1970) and then 60 percent (1970–1980). Although the general decline in commodity prices hurt the mining industry in latter years, employment levels have been maintained, actually increasing 9.6 percent in the period 1965–1975 (SPP, 1980: vol. 3. Table II.1.17). Manufacturing activity has increased substantially, largely as a result of its being favored by the investment policies of the Echeverría administration (1970–1976) and its being granted in-bond (*maquiladora*) status during that period. The industrial zones to the south of the city grew rapidly during the period 1960–1980, as did most economic indicators. Car ownership increased 78 percent in the years 1975–1979 alone.

The only other city that shows a marked difference when you compare our livability scale with Unikel's *"nivel de vida"* and Conroy et al.'s "socioeconomic opportunity index" is Mérida, which ranks low on our scale and third on both the others. The reasons are clear from examining Table 8. Mérida ranks high on the rate at which people are paid wages below the minimum salary (21%) and lowest on the percentage of houses that have sewer connections, the two variables that are weighted the heaviest in our "livability" scale. Neither appears in Unikel's scale, and in Conroy's scale only the associated item "piped water available" appears. From our experience with the National Institute of Community Development, we would argue that both these items are extremely important in determining how attractive a city is. Both are intimately associated with sustaining a decent life. The lack of sewer connections has drastic effects on health, is dangerous to the water supply, and is responsible to an important degree for the devastating gastroenteric infections that raise child mortality so high and impair the health of adults, their productivity, and their ability to enjoy life.

Being paid subpar wages speaks for itself. If the minimum wage is insufficient to sustain life in anything like socioculturally decent conditions, how much worse for people who make less than the minimum wage. It is not just that their diet would be meatless and monotonous but that it would lack sufficient vegetable protein as well and would require the expenditure of fully 70 percent of the household's income to buy. That would, of course, mean that no visits to doctors could be even entertained, and even the small but nevertheless palpable expenses for postprimary education would make this beyond the family's means.

Conclusion

This analysis has shown that some cities during the study period were more attractive than others. Mexico City seemed that way to many Oaxacans, some of whom were apologetic about their lack of "get up and go" by staying in Oaxaca in relative poverty and bypassing the opportunities for better jobs that existed in the capital. That feeling no longer exists, as of 1987. In our 1987 interviews of 30 households in the city of Oaxaca, not one evidenced a desire to go to Mexico City; in fact, a number of them had returned from the metropolitan zone to Oaxaca.

But if you were an ordinary Mexican family looking for the best break you were likely to get in a Mexican city, San Luis Potosí would be your best bet. The nation's capital was second best in 1976–1979, but is no longer as attractive, since it has been the hardest hit by the economic crisis, harboring, as it always has, the majority of the industrial and commercial activity of the country. In addition, the metropolitan zone is running out of places to live; ordinary people who are looking for housesites for their children are being compelled to go as far as Chimalhuacán and even Texcoco, commuting by bus and metro 10 miles and more to the center of the city and more than two and a half hours to the northern industrial zones in Naucalpan and Azcapotzalco.

Tampico looks attractive, but not if you are thinking of moving there, unless you are already a member of the oil workers' union. The same is true for Reynosa, which despite the presence of an oil industry is ranked very low. Oaxaca still lacks any industrial base and remains an example of a city that survives by ordinary people taking in each other's washing. The number of domestic servants in Oaxaca is still very high for a country as developed as Mexico, and as late as 1987 some domestic workers were earning as little as 20,000 pesos or US $14.50 a month.

We conclude with the thought with which we started; namely, that although the cities of Mexico are different in appearance, style, geography, history, and even dialect, still, for the majority of Mexicans, ordinary Mexicans, life in the cities is a relatively similar challenge from city to city. In the next chapters we shall see how ordinary Mexican families meet that challenge.[11]

PART TWO

The Mexican Urban Household

In this section, consisting of chapters 3, 4, and 5, we discuss the anthropology of the Mexican household (chap. 3), the household as the locus of production and reproduction, and how this plays out in the lives of ordinary people, relying heavily on our field work. But before we take on this task, we review some important concepts that have influenced the discussion, principally those that have attempted to embed those activities that we call the household in a theoretical context. This leads us to discuss the work of Gary Becker and his associates (and antagonists), the economic theory of fertility, and the rationality of household organization. We next discuss the literature on "survival strategies" to see whether the notions of rationality and decision are applicable to Mexican urban households. Last, we talk about the bargain between the household, the individual, and the state and describe the struggle among these three social actors. We see the household as one actor in a struggle with its own members, on the one hand, and the state, on the other. The household is taken to represent the collective interests of the people who live in it, and this is of course not the same as the interests of the individual members. The politics of the household are based on a state-endorsed ideology that gives advantages to males and to seniors in the household. Many of the sacrifices that are made by the household are made by women and children. The state is seen as having interests in the household, which it exploits for the purposes of labor supply. The state's interests lie in furnishing itself with an ample supply of educated, trained, and preferably docile and obedient workers who are willing to accept very low pay for their work in industry or casual jobs. In this way it can furnish itself with the subsidies that will allow it to survive in a competitive world despite the inefficiencies of its economy that derive from dependency and the lack of scale of a home market. Ordinary people pay those subsidies in the form of sacrifices in their standard of living, their dignity, and their feelings of security. They try to defend themselves with their families, homes, and households. In chapter 4 we study the "sociology of the household" by which we mean "household composition," organization, and living conditions. We construct a simple typology of households based on the underlying family types of singletons, matrifocal families, nuclear families, and complex (mainly lineally extended) families. In chapter 5 we show that families that are better off are larger, with more children, and more advantageous dependency ratios. This leads us to a discussion of the trajectory of the household through the domestic cycle.

3. Households, Strategies, and the Economic System

Before we embark on a discussion of the Mexican urban household, there are some definitions and theoretical problems that require discussion. Some are definitional, but others go to the heart of the problem of what households and families are for, and what they do, and how they are composed, and some involve a critique of the theoretical language that is used to describe households and families in their social and economic context. First, the basic definitions.

Family and Household

In this study we do not examine families directly. We look at *households*. The household can be defined, as Schmink (1984:89) has done, as a "coresident group of persons who share most aspects of consumption drawing on and allocating a common pool of resources (including labor) to ensure their material reproduction." It is not a family. For our purposes "family" is a cultural category, while "household" is an analytic category. Briefly, to obviate a literature that is well covered in Buchler and Selby (1968), Schneider (1968), Selby (1971), and Yanagisako (1979), this means that "family" is a culturally salient term about which people talk, think, fight, and devise ideologies, whereas "household" is a term that such analysts as census takers use. The distinction is arbitrary, but it is also useful, for when we talk about "family" we are talking about a cultural concept, while "household" is an analytic construct. Perhaps unfortunately, the distinction is not nearly so clear in Spanish as it is in English. In that language *hogar* means "household," but it also means "family life" and is taken to be synonymous with *familia*. For this reason Latin American social anthropologists and sociologists use the term *grupo doméstico*, or *unidad doméstica* to coincide with "household," for these terms are clearly analytic constructs and not cultural concepts.

The Importance of the Household

Households are defined by the kinds of families (or absence thereof) that compose the residential group. Just about three-quarters of the households in our study are made up of nuclear families. Defining households as based on underlying family types, that is, as single-tons, matrifocal (father absent, female-headed families), nuclear, and complex (a residual category meaning "greater than nuclear") yields the following distribution (see Table 12).

The way we collected our data influenced the distribution of household-family types. If a house stood alone in a housesite (*solar*), we interviewed the head or head's spouse. If there was more than one house, or if we had entered an apartment house (and often it is difficult to tell from the street), we chose one of the houses or dwelling units at random. We defined the people who lived in each dwelling unit as a household by asking the question, "Please tell us the names of the people who live here in the house." As a result, we have underestimated the complexity of household organization, since a broader definition might well have included people who lived in the other houses, who were almost certainly related. On the other hand, we have overestimated the number of singleton or matrifocal households because we were not careful in the first phase of the project (when we interviewed in Oaxaca, Venustiano Carranza, Mérida, Villahermosa, and Reynosa) to inquire about absent spouses and household members. Because we concentrated on the units themselves and did not inquire at all deeply into relationships among households, we cannot discuss to what degree households effectively extend to relatives across the street or even back to the village. We are unable to contribute to the discussion of the importance and extensiveness of *"redes de seguridad,"* or networks of kinship

Table 12. *Frequency of Household Types*

Household Types	Number	Percentage
Singleton	226	2.4
Matrifocal	684	7.2
Nuclear	6,929	73.2
Complex	1,620	17.1
Total	9,459	

and exchange, that are so important in understanding the urban family in Mexico, as Alonso (1980), Arias (1982), Arizpe (1982), González de la Rocha (1986), Kemper (1981), Logan (1981), Lomnitz (1977), Royce (1981), and Vélez-Ibáñez (1983) have done.[1]

On the other hand, we can feel very confident in asserting that practically everyone in Mexico lives in family-based households. Table 12 shows that only 2.4 percent of the nearly 10,000 households that were interviewed were singletons. And in approximately half of these (details in next chapter), spouses were reported only temporarily absent. Matrifocal households made up just over 7 percent of the total in our sample, which is quite low compared to the United States, where among some social sectors matrifocality is the rule, reaching 27 percent for all families in 1987: 44.3 percent for black families and 25.7 percent for white.[2] We feel safe in asserting that the household, normally consisting of a nuclear family, is *the* important context of experience for every Mexican. Development, change, near catastrophe, poverty, and sometimes grinding misery have not destroyed the Mexican family as it seems to have in the U.S. American underclass, if Auletta (1983), Moynihan (1986), and Wilson (1987) are reliable guides.

Household Production and Household Firms

A second set of considerations has to do with how households are constructed and what their purposes are. We organize the discussion about how they are constructed by suggesting that they are socially and ideologically constructed on the basis of cultural constructs of gender, power, love, and obligation as these relate to the families that constitute them. We address the question of what households do by discussing the literature on the household as a rational decision-making production unit. The first discussion is about Mexican ideas and Mexican culture, and the second is about analytical mechanics, deduction, and hypothesis testing.

The Social (Ideological) Construction of the Family

Perhaps the first thing to affirm is that neither the family nor the household is a "natural unit." It would seem unnecessary to assert this, but as Harris has pointed out, the term "natural" is profoundly rooted in European thought, is associated with women, and has negative associations of great antiquity as well as contemporary power (de Beauvoir, 1972). The nuclear family may be close to being universal (Buchler and Selby, 1968: chap. 2), but as Laslett (1983) and the members of the Cambridge Demographic Group have shown us, it is a concrete historical form that has to be understood that way.

The family is an ideological construction based principally on the notion of *respeto* and the gender (and power-related) concepts[3] of "macho" (and associated terms), and *abnegación* and instantiated in kinship and kinshiplike terms, with a presumed relationship of common blood, or "spirit" to define its boundaries. The family is the boundary of trust of "*confianza,*" since one cannot really trust people outside the family. Put another way, if people inside your family deceive you then it is an act of treachery. It is difficult for foreigners who are not native speakers to carry out a cultural analysis of these terms, and the best approach is to describe the working of the ideal family in order to see how these concepts work out to define the household/family.

The Ideal Family and Household

The ideal household[4] works in this way. It is organized around an authoritative, hard-working, permanently employed, nondrinking father, handing over his wage packet to a self-abnegating, altruistic, suffering mother, whose children, seeing her sufferings and appreciating the nobility of the paternal example, help in the household from the earliest age (girls earlier than boys) and contribute their earnings to the household budget by handing them over to their mother from the age of eight to 12, retaining nothing for themselves.

The organization of work under ordinary conditions of capitalism favors this kind of a moral ordering in the family by giving preferential employment status, pay, and prestige to the male worker, but denying him enough wages to support his family. Economic cooperation within the household is thereby enjoined and, ideologically, endorsed. The deficit in the domestic budget must be made up by cooperative enterprise, either in subsistence production, in casual work, or in waged work on the part of other family workers. The work of women and children is seen as complementary to and therefore less well paid than the work of the (normally) male head of the household. This is strikingly reminiscent of the ideal working-class family descriptions of working-class London at the turn of the century, when workers there were being drawn into the fully proletarianized status with about the same historical depth as many Mexican urban families. Davin (1984) has described the working-class family in turn-of-the-century London in a way that rings true of ordinary urban Mexican families. The suffering mother was a central symbol of family life. "Old people often recall how their mothers would stint themselves: 'Oh, my mother went without for us, yes, yes. I've known her to wipe the plate round—with a drop of gravy—and tell my father she's had her dinner' (interview pertaining to turn-of-the-century London). They remember too how endlessly their mothers worked and how rarely they rested, or treated themselves to anything requiring money. The mother's continual labor and self-sacrifice often inspired fierce loyalty and protectiveness in the children, and great eagerness to do whatever they could to help, especially by earning" (Davin, 1984:227).

In the ideal household, children serve the interests of the parents by being obedient and by taking into account all collective needs and goals before individual ones (nicely meeting the assumptions of a single household utility function of the "new household economics"). They are as abnegating in their way as the mother is in hers. They bend to the authority of their father. They hand over all their earnings to the mother, who will give them back money for their *gastos* (expenses) for transport, books, snacks, clothing, and (very occasional) entertainment, as well as courting expenses if they are in the right place in the marrying queue, that is, if their elder siblings are either all married or have tried and failed the courting challenge.

The household budget will be allocated with a priority placed on collective expenses first. Moneys will be invested by the household in maintenance and expenses and then for housing, extra food, educational expenses, and festival occasions. The costs of new housing or improvements will be met through "windfall income," as Musgrove (1978) has suggested, but family maintenance expenses are met out of family income. Children will begin contributing to household welfare by the time they are five or six, with girls tending more to contribute unpaid labor to household subsistence and maintenance and boys tending to scrounge for errand and casual jobs in the street. By the time children are 12, their paid and unpaid contributions are substantial, with girls earning less than boys but contributing a larger proportion to the household budget.[5] We guess that, in the majority of cases, a permanently employed child has repaid all the expenses of his rearing by the time he is about 20. The ideal child will have done that by paying all his education expenses through the preparatory level, have married, taken up residence with his parents, and reimbursed them for their expenses for his raising.

In the ideal family, there is concord and respect shown to all. Obviously children respect and obey their parents and join with them in considering the needs of the household before their own selfish needs, and similarly parents respect their children, giving them the privacy they need in cramped quarters for courting and pursuing their lives and their studies, and not pressuring them unduly for monetary contributions and for work around the house. Boys are favored more than girls in domestic tasks, for it is natural (*lógico*) for girls to do housework, but not so for boys. Girls respect their brothers, a respect that is tinged with obedience; and boys respect their sisters and are bound to protect them from a sexually exploitative world.[6]

The realm of women is the interior of the house and patio, while the realm of men is outside. Women's duty is to create and sustain the *hogar*, and they are the focal warmth of this loving refuge. In the ideal family, marriage of the children does not create tensions among them, for there is a marriage queue by age. Each child in turn has a preferential call on what family resources are available, unless he or she fails the "courting test," which usually is indexed by boys beginning to drink and engage in promiscuous behavior, thereby signaling that they are throwing away their chances, and in a similar vein by girls losing their virginity and their *novio* (boyfriend). Residential arrangements are discussed by parents, and the couple is more or less assigned to one of their residences. In the old days, there was a preference for virilocality, so that the couple would live with the boy's parents, but nowadays that no longer holds. It is the needs of the older generation and the opportunities provided for the younger that weigh heavily in the scale of residential choice. Couples will want to have their own residence, but not until a couple of children have been born.

Some Consequences of Family and Household Ideology

The idealized picture of family life omits to consider, as women and children in particular remind us, that this moral ordering is not without costs. The ideal assumes that the interests of the head of the household or at least the interests of the adults in the household are the same as those of the children, and this is simply not so. The household, as González de la Rocha (1986) explicitly, repeatedly, and rightly insists, is a category and a social field that is replete with contradictions. It is the locus of reproduction of gender ideologies and of the ideologies of subordination to authority. The legitimization of the presidential system (Carpizo, 1978; Scherer, 1986), the social pyramid of Mexican society (Paz, 1985), and the *autoridad* that orders all levels of Mexican society is both recapitulated and instantiated in the ideology of the family-household. Worse, from a democratic-liberationist point of view, the arbitrariness and vio-

lence that are so fundamental a part of the exercise of authority in
the social and political system, find expression and legitimization in
family violence, most often by men against women. The interests of
all members of the household are not the same, and as in the political society, there are no sanctioned means for the mediation of interests except through the power of decision of the father, the interces-
sion of the mother, and such persuasive power as she can bring to
the process.

Children have two major goals that are antagonistic to the collective moral philosophy of the ideal family. They want to get ahead,
and they want to get married. And they need independence to be able
to do either. Most newly married Mexicans do not mind living with
one set of parents for a spell, but they do not wish to do it for more
than a few years unless the relationship between the newly married
couple and the family is unusually warm and free of prying eyes,
censorious judgments, and psychological pressure. Mexico is like
other capitalist countries in that much emphasis is placed upon notions of personal liberty, freedom of choice, and individual responsibility and autonomy, and the children are as affected by this state-
sponsored ideology as are the adults. Perhaps more so, since they are
more highly addicted to the products of the U.S. media and, at least in
the poorer barrios of Mexico City, report that such U.S. productions
as *The Dukes of Hazzard* are what they like to watch on television.
The children understand the paradox that they can only perfectly
serve the families they are having by renouncing their responsibilities to the families from which they spring, at the same time as they
embrace the ideal family ideology which has little to say about splitting a family and much to say about raising one, once it has split. On
a less lofty plane, living with one's parents or parents-in-law puts severe limitations upon what one can do and how one can behave in
the expressive sphere. For the husband, drinking, sexual adventur-
ings, and extravagance in the personal and economic spheres are curtailed by living with one's parents, who demand the evidence of
respect as shown by discreet behavior. For the daughter-in-law, freedom of movement and her ability to commandeer the undivided loyalty of her husband, as well as her freedom of decision, are curtailed;
and often enough parents-in-law forget that she is not a daughter and
impose restrictions of an even more galling sort, symbolized by im-
pertinent queries about her goings and comings and suggestions and
implied criticisms about the way in which she spends (read "wastes")
her husband's earnings.

The ideology of the ideal household does not take into account the problems arising from the ambitions of the children to get the jobs for which they are eligible if they have higher educational credentials. Jobs which remunerate education are very scarce in the provincial cities of Mexico, and if the young man has managed to educate himself through the preparatory level, as 14 percent of our sample of heads of households has, then it may well be necessary to move to Mexico City, Monterrey, Guadalajara, or the "other side" (U.S.) in order to secure employment that rewards the educational level for which one has worked. And only in the capital can one find those most desirable of employment opportunities: livings which are guaranteed and where effort is minimal, that is, jobs in the bureaucracy.

The family is all-important in Mexico, ideally and in fact. Our research on the households that send migrants to the United States has led us to conclude (Selby and Murphy, 1982) that the vast majority indeed do send them in order to preserve their own integrity. People do not leave their families of orientation in order to break them up. And one should be careful with terms like "independent family" or "splitting" or "break up" of the family because often the reality is much less dramatically rending than the vocabulary. A son or daughter may establish a residence in the same neighborhood, or even the same block, and the consequent closeness, visiting, and sharing can approximate extended family living. Lomnitz (1977) has documented how poor neighborhoods in Mexico City can establish networks of kin, friends, and fictive kin if the people living there have or create the basis for strong and enduring social relations. In the case of the poor people she studied, the members of the *colonia* came from the same village and were recruited to the same occupation by relatives who had established themselves in the city. Networks of kinfolk from the village were reconstituting themselves in the capital, as have kin from other villages, like Tzintzuntzán (Kemper, 1981), while Nutini[7] has similarly found Tlaxcalán villages reconstituted in one of the most densely populated areas of Mexico City, La Viga, where they have constructed ceremonial organizations and effective kin networks. It is not clear how frequently this occurs, and there may be special conditions that facilitate the process which do not occur ordinarily in the process of urbanization. It is probably rare that neighborhoods are built up of related families, re-creating villages within cities, and even when it does occur the life span of these villages within the city is probably quite short. But their existence, along with the data on networks of security mentioned above, warns us not to make a too catastrophic interpretation of the break-up of the household, which in fact may be a gradual process and an intrinsic part of the domestic cycle of maturing households, largely controlled and constrained by economic factors: the availability of house lots in the parental neighborhoods and geographical job mobility on the part of the children. The Mexican household-family is extraordinarily tenacious and resilient, as it has to be in order to meet the challenges posed by an exploitative and predatory society bent on extracting surplus from it sufficient to assure national survival in a competitive world.

The most critical point of domestic cycle in the city occurs when children leave school, establish themselves in some kind of occupation or more or less steady employment, and begin to wind down the courtship period in preparation for marriage. As we shall show in chapters 5 and 7, it is critical at this point that the parents of the couple make arrangements to retain the couple's services in both the wage-earning and the domestic sphere. Although the ideal family favors the retention of the married son, in fact, married daughters are almost as common. In the case that the bride's mother is living alone in a substantial house, visiting and contributory arrangements will be set up so that the couple can live with the bride's mother. The couple visits the husband's family on a regular basis, and the husband contributes to the maintenance of his natal household, and, at an absolute minimum, provides appropriate moral guidance for the younger brothers and sisters. Material assistance is welcomed and expected as well. During the period of early marriage, the couple will be saving to establish their own household, and the length of time that they will stay with their parents depends on the degree to which they feel that they can save money for their future, that is, the degree to which the material demands of the extended household make sense in terms of their own budget, as well as the degree to which they can maintain some kind of autonomy, all within the moral constraints of family living which accord respect to the parents.

All people struggle with their ideologies, and since the ideology of the family and the household favors adult males, it is not surprising to find that women and children pay most of the unrecognized costs of national economic survival or development. In this section we have tried to give some idea of the dialectic that defines the field that we call the ordinary Mexican household in its ideological and practical moment. In the next section we turn to the question, "What do households do, and how do they operate?" We begin by consider- ing the ordinary Mexican urban household as the locus of production.

The Household as Locus of Production

Anthropologists studying rural villages, including Mexican villages, have experienced little difficulty in identifying the village household as the locus of production and consumption. Households are labor collectives, made up mainly of kinfolk engaged in subsistence, milpa agriculture, and that would seem to qualify them for the most important units of production and consumption. They are certainly not the only units engaged in production and consumption. Cooperatives that consist of neighbors, kin, and fictive relations are certainly important in production, and sometimes the village as a whole engages in productive activities, as in the *parcela escolar,* or the parcel of land that is set aside for the school.

Subsistence products are not all that households produce. Petty commodity production may well be important, as may waged work. And it may be useful to take a broader view of production as we did in our work on the production of values in villages and towns (Selby and Hendrix, 1976; Murphy, 1973; Stepick, 1974). Our model of efficient village producers had them producing unlikely (and not easily measured) goods, such as self-respect, the respect of other members of the community, and social and psychological security.

But often we think of the urban household as a unit of consumption only and neglect its productive activities. This is a mistake, even though in the city the production possibilities of the household are somewhat more limited. According to José Alonso (1981), 10 percent of the adult women in three *colonias* of Ciudad Netzahualcoyotl were engaged in *maquila,* or household production of intermediate textiles on subcontract.[8] Jorge Alonso (1980:160) has reckoned that in the Colonia Ajusco (Mexico City) 34 percent of the work force was engaged in activities (as artisans, *maquileros,* construction work, storekeeping, or street vending) that one could regard as productive work either partially or wholly based in the household. Our observations in the cities of Mexico have permitted us to see a large increase in the amount of productive work that is going on in the home as a result of the decrease in formal sector employment. In particular, in 1987 in Oaxaca, 28 percent of the sample of 604 households were recorded as being the site of remunerated economic activity, most of it commercial, or in petty commodity production. As we shall see in greater detail in chapter 9, the increase in informal sector productive work by women has been very dramatic in Oaxaca from 1977 to 1987 and, we suspect, for the other cities of Mexico.

So it should not be thought that households are not the loci of production by men, women, and children. And as Jorge Alonso (1980), González de la Rocha (1986), and others have shown, the household is the production site that yields the greatest proportion of value added to formal sector enterprises, under conditions that best remind one of Engels' descriptions of the domestic enterprises of the working classes in midnineteenth-century England (Engels, 1952; González de la Rocha, 1986:174–178).[9]

There are other ways of thinking about household production, more in the line of our earlier thinking about the production of values, and one of the most important recent ones has been called the "new household economics," which is associated with the work of Gary Becker at the University of Chicago. This is a useful tradition to study, if only briefly, because it will explicitly involve us in the discussion of the household as a decision-making unit, which is a topic that bears full scrutiny for both its theoretical implications (asking the question about whether the language of decision theory is applicable to the situation) and because of its ethnographic implications (to what degree can notions like "decision" and "rationality" be said to apply to the behavior of the household).

The "New Household Economics"

The most interesting attempt to extend the neoclassical microeconomic framework to the family has been that of Gary Becker (1964, 1976, 1981) and his associates. Their work is worth discussion, particularly as we have adopted it in our understanding of changing fertility and natality patterns in urban Mexico, as well as in our understanding of household composition. We believe that the "economic theory of fertility," as it is called, has some relevance to urban Mexico but that the general theory of the household as maximizing firm is far too restrictive.

In the theory of the household production functions, Becker assumes that "time and goods are inputs into the production of 'commodities,' which directly provide utility. These commodities cannot be purchased in the marketplace but are produced as well as consumed by households making market purchases, own time, and various environmental inputs. These commodities include children, prestige and esteem, health, altruism, envy and pleasures of the senses" (Becker, 1981:7–8).

The most important good produced by families (or households) is children, and the value of children is determined by the availability of birth control techniques, as well as the obvious biological constraints. The number of children desired is determined by utility considerations in the context of a household's desire for a full array of goods, subject to budget constraints. Children cost money and compete for income, but more important, they compete for the time of the parents, particularly the mother's time. Considering the value of children alone (i.e., apart from the choice between children and other consumer goods), the number of children desired will depend upon the opportunity costs associated with their nurturance (i.e., the market value of the time that is lost by having children, particularly by the mother's absence from the labor market) and the tastes of the family for what Becker calls the "quality" of the children. "Quality" is an elusive concept because it refers to the attributes of children that enter into the utility functions of parents. But more practically it has to do with the health, earnings, education, and wealth of children. And these are determined, in part, by the investments made in them. There is a trade-off between having many children and investing little in them (the "quantity option") or having few children and investing much in them (the "quality option"). The expense of rearing children is offset by their capacity for contributing to the household's economy, and this is why households in farming communities have more children than urban households.[10] The number of children, given fixed preferences for quality/quantity, will be influenced by the income-earning opportunities available to them, as seems to be true in urban Mexico in the present period of economic crisis.[11] (This has been an important influence on the reduction of the urban fertility rates in the past 15 years, particularly during the years of the economic crisis in Mexico.)

The "new household economics" is interested in more than the economic theory of fertility. Among the many topics that have interested sociologists and anthropologists, two are important for our discussion because they bear on the organization of the household: "altruism" in the family and the sexual division of labor. Our uneasiness about the treatment of these topics in the theory, as well as our uneasiness about treating the household as solely a production unit, will induce us to broaden our theoretical discussion to that of the household as a reproductive (in the broad sense) unit, as well as a productive one.

Altruism (Becker, 1981: 172–176) has long been recognized as important in family organization. Becker quotes Adam Smith (1976: 321): "Every man feels his own pleasures and own pains more sensibly than those of other people. . . . After himself, the members of his own family, those who usually live in the same house with him, his parents, his children, his brothers and sisters, are naturally the objects of his warmest affections. They are naturally and usually the persons upon whose happiness or misery his conduct must have the greatest influence." In particular, the husband's utility depends on that of the wife, and the husband is induced to spend his income on the wife because he gains utility from her consumption as well as from his. In this sense, just as in the biological sense of the "selfish gene," the altruist is not altruistic in the ordinary language sense, but selfish. He gains from his sacrifice and loses from pure selfishness.

One of the more interesting characters in this household of stylized facts is the Rotten Kid. Becker explains (1986:283):

> Consider a parent who is altruistic toward her two children, Tom and Jane, and spends say $200 on each. Suppose Tom can take an action that benefits him by $50, but would harm Jane by $100. A selfish Tom would appear to take that action if his responsibility for the changed circumstances of Jane's were to go undetected (and hence not punished). However the head's utility would be reduced by Tom's action because family income would be reduced by $50. If altruism is a 'superior good,' the head will reduce the utility of each beneficiary when her own utility is reduced. Therefore, should Tom take this action, she would reduce her gift to him from $200 to less than $150, and raise her gift to Jane to less than $300. As a result Tom would be made worse off by his actions.

Children still cooperate then only as long as the head of the household has what Hirshleifer (1977) has called "the last word." It does not matter whether the head knows that Tom is being nasty to Jane or not, he or she only has to know about the changed condition of the household in terms of diminished utility and to act rationally to change the structure of rewards to prevent this deterioration. As Becker has stated (1986:282), this theorem (the "Rotten Kid theorem") does not rule out conflict in families and can readily accommodate facts of family existence like sibling rivalry (but see below for our reservations).

The other topic of interest developed in the theory apart from the fertility rationale and altruism is the sexual division of labor. Briefly, the division of labor in society is explained by differential returns to different individual endowments and return to human capital investments. The most efficient family is one in which each member devotes his or her time to those activities to which he or she is best fitted by these two criteria. However, when economic forces external to the family change the expected gains from investment in human capital and raise the opportunity costs of a given division of labor, then it will change. And this is precisely what has happened in the United States and Mexico in the past decades. In the United States, as Mincer (1962) has shown, the substitution of market work for work in the home was caused by the increased potential earnings of women. The associated rise in the costs of bearing and nurturing children has, in equal measure, reduced the birth rates in these two countries.

Objections to the Theory for Understanding Households (in Urban Mexico)

Although we have found the economic theory of fertility and the household useful and provocative for our work, and although we have been influenced by many other writers in the tradition, including those of a more historical-sociological bent like Easterlin (1968, 1973, 1978), there are a number of objections to the theory for understanding households in Mexico that need to be raised.

We do find ourselves objecting to and imprisoned by the strict version of the assumptions of the kind of utility theory that Becker is using (utility maximization with fixed preferences and the assumption of perfect information in the composition phase of the household via marriage). So far as preferences are concerned, they obviously change, and that preferences for children change with parity is well known as in the studies by Namboodiri (1974) and Simon (1975) for the United States, Park (1978) for South Korea, and Bulatao (1981) for a comparison of the United States, South Korea, and the Philippines.[12] Although we certainly agree that households are productive in both the material and nonmaterial sense, we are not at all sure that they really optimize. It is not necessary to review the general criticisms of utility maximization, since they are well known and can be reviewed in Liebenstein (1976), Lesbourne (1977), and Simon (1978). But one may sensibly ask for the empirical data that would lend support to such a proposition or for the theoretical consequences that would lead one to assume it in the first place. And this is lacking.

In Becker's case the assumptions are motivated by the reasonable desire to furnish a deductive theory of the family and are justified by the proposition that only efficient families survive in a competitive market and that "efficient" is the same as "optimal." But this is not true, since survival requires reproduction not optimality. Being efficient enough to reproduce is not the same as being as efficient as one can be.

The perfect information assumption is also quite dubious, since it assumes that household members will share information about each other. An important aspect of our argument concerns the competition among household members for limited investments and opportunities. Individual strategies would seem to indicate the free exchange of information only in those rare families where all are for one and one for all, whereas under conditions of poverty and (sometimes) violence, a Hobbesian condition would seem to be a more appropriate descriptor. This condition similarly questions the notion of a single household utility function (Nerlove, 1974; Sawhill, 1977) with a household head conveniently representing the interests of all. The Rotten Kid theorem shows that the putative head need not be aware of the tensions, shirkings, and deceits within the family, so long as he or she has the "last word," but what if the head is a supposititious notion in the first place. Even though we, like most interviewers in Mexican survey work, use the term *"jefe(a) de la casa"* to identify a cultural category that is recognized by ordinary Mexicans, we should not be fooled into believing that Mexican households actually have effective administrative heads with full powers of command and execution. Quite the opposite is the case. As our observations and others' have shown, households where the headship is lightly and cautiously held, where machismo is held tightly in check and where negotiation and generosity toward the people who are not the head, are those where efficiency or optimality are attained. Our view coincides with that of Lewis (1963) and González de la Rocha (1986) that the household is a site of conflict, a locus of contradictions, and a field of negotiations without a single decision-making household head. The family of *Children of Sánchez* provides a very good example of why the "single head assumption" and "efficiency and optimality" are contradictory in the urban Mexican world. Jesús, the father, and all his children agreed in their testimony and demonstrated in their behavior that their family was a failure. Jesús lamented that, even though his ambition to build a house for his children to live with him potentially could be realized, as it subsequently was in Ixtapalapa (Federal District), his children would never come to live with him as he wished them to. And they never did. He realized that he had never succeeded in keeping

his family together as a loving, caring kin who respected each other.[13]
They were precisely the opposite of Sra. Concepción's family, which
we have taken as our example of one important type of efficient family. And as the children so often complained, Jesús was the feared
head of the family and chief executive officer par excellence. Theirs
was a family in which little negotiation took place. Becker's theory
tries to get around the restrictions of the single decision maker/
single household utility function assumption by suggesting that
there is no need for the household head to know about selfish shirk-
ing in the family, but as Yoram ben Porath (1982:53) points out, the
kind of monitoring of members' conduct that is required involves invasions of privacy, and these invasive practices may well lead to diseconomies of scale (in large families where there is a need for much
"checking up") to the point that the families must remain small. But
monitoring and checking, as in the Sánchez case, lead to deceit and
the withholding of information and thus pose contradictions to the
underlying assumptions of the theory as well.

Much has been written on Becker's theory. One critic has concluded, as we have, that Becker's (1981) book *A Treatise on the Family* is a major achievement. It is the most important book on the
family to appear in many years. But he concludes his review and its
associated work with the following comment, with which we find
ourselves in agreement:

> Becker's theories both understate and overstate the role of
> culture and institutional arrangements. They understate the
> impact of the family by denying that institutions have any
> independent effect on behavior. They overstate the importance of
> social structure by leaving moot the questions of the origin of
> the driving forces. . . . The time seems ripe to work towards
> models that acknowledge the coupling of individual decision-
> making and institutional change. Such models would recognize
> that individuals can rebel against norms and coercion, and acting
> collectively can change institutional structures. They would also
> recognize that institutional structures have lives of their own,
> that they impart inertia to social systems, slowing the speed of
> response to changed conditions.

Decisions and Survival Strategies

In this study we employ the concept of "strategies of survival" but
with reservations that bear consideration. The first explicit use of

the term *estrategias de supervivencia* was in the work of Duque and Pastrana (1973), who used it to study the struggles of poor families in Santiago de Chile. And as Omar Argüello (1981 : 192 n. 1) has pointed out, the use of strategy was an "objective" or imputed. In the fashion of William Geoghegan (1969) in social anthropology, life course events were studied, and the trajectory of related events was called a "strategy" whether the people in the households themselves were aware of making some kind of decision or not. Duque and Pastrana examined the way in which the family organized itself to take advantage of whatever income-earning opportunities there were in Lima in order to assure its material survival. The concept of "survival strategy" was adopted by the Program of Research on Population in Latin America (PISPAL, 1978) for the study of demographic events. Susan Torrado (1978, 1981) working in PISPAL, found the notion useful but restrictive and defined two concepts: "survival strategy" and "family life strategy." The first referred more narrowly to the demographic and biological conditions of material survival, while "family life strategies" was a broader notion that included (Torrado, 1981 : 227–229) the demography of the household, socialization and teaching of the children, organization of family consumption, the choice of living site, the choice of outside family members, cooperation with households outside the immediate one, domestic cycle, migration of family members, and preservation and maintenance of the household in general.

As the concept was used, it became clear that there were questions as to its applicability. Was it so general as to include the "survival strategies" of the rich as well, as they struggled to keep family firms together and to win political office (the Kennedys were instanced)? No; the rich and the powerful did not seem to be well described as "surviving." Was it proper to talk about "survival strategies" as applicable to both good economic times and bad? Or as Valdés and Acuña (1981) suggested, was it better to reserve it for periods when conditions were deteriorating and survival was truly an issue? No; times were bad enough for poor people in the cities of Latin America to assure the relevance of the term at any time. Torrado's further query (1981, 1982) was, To what degree does the notion of "strategy" imply rational actors and notions of rationality, and what criteria should and could be adduced to define rationality? Rationality considerations invite us to integrate Becker's framework into the analysis, with the difficulties suggested above. If one is using rationality metaphorically, then obviously one is using the term in the sense of "bounded rationality," which introduces a whole new set of difficulties having to do with whether such notions as bounded rationality or satisficing and their congeners help at all. In Mexican cities, to be blunt, many of the poor would be better off to emigrate to the United States, and it seems gratuitously condescending to speak of their implementation of bounded or restricted rationality on site. Is "survival strategy" a universal notion, or is it to be confined to determinate historical periods and certain kinds of society defined by well-defined "development styles"? Both Argüello and Torrado agree that it should be made clear to what social sectors the concept refers, since it is always used to describe the struggles of subordinated groups who are marginally inserted in the productive apparatus of their society. For this reason Argüello defines "survival strategy" as "the set of economic, social, cultural and demographic activities utilized by the strata of society who neither possess the means of production nor are fully incorporated into the work force because they cannot obtain regular incomes to maintain a given level of living." He points out that this definition will do quite nicely for peasant sectors in the process of *descampesinización* (losing their land and being forced to find alternate means of support outside subsistence agriculture) as well as for urban worker families in the informal sector.

A further set of problems has to do with the analytic units involved in the strategizing; some had decried the use of the term "family" as a universal concept at all, preferring instead to talk only about concrete operating household groups, or *unidades domésticas*. But even then it is hard to know where the boundaries of these concrete groups lie: in our case do we decide that households are contained and defined by the *barda?* Or do we go further as Alonso (1980), Arias (1982), Arizpe (1982), González de la Rocha (1986), Kemper (1981), Logan (1981), Lomnitz (1977), Royce (1981), Torrado (1981, 1982), and Vélez-Ibáñez (1983) have done and include outsiders, neighbors, distant relatives, and even associates back in the village as effective members of the household collective?

In this we agree with Rodríguez (1981) that the family/household is best defined for concrete social circumstances and taken to be that arena in which strategies, such as they are, are organized. This at least takes us away from the notion that individuals make decisions apart from family, kin, and household considerations. We further agree with him (1981:234) that one has to take account of four factors in understanding the situation of the household: characteristics of the labor market, the characteristics of consumption, the characteristics of the actions of the state, and the living conditions of the population.

Conclusion: Decisions, Strategies, and Survival

In summary there are three major problems which have to be handled in the discussion of households and how they get on in urban Mexico. The first problem is the *consciousness* issue. To what degree can we say that a decision or a string of decisions that we call a strategy exists if the people who put it into effect are not aware of it? No one is aware of all the implications of a decision, and unintended consequences of decisions are the staple of social science research. But to what degree can we say that having a child is a decision and is part of a related string of decisions with concrete outcomes to be called a strategy of survival? It is hard to say and remains an empirical question. Sra. Hernández, who said triumphally, "With these children I am going to win," certainly realizes she has hit on a winning strategy, although it may not have been clear to her at the time she was alone and raising her five small children. But later on when we analyze survey data, it is less clear whether we may impute the notion of "strategy" and "decision" to people whom we did not explicitly interview on the topic. In other words much care has to be exercised in our interpretation of the notion of "decision" and "strategy" so as not to suggest that we are taking the "consciousness issue" for granted.

Second is the *density of alternatives* issue. If we are using the term "decision" in a proper fashion, we are assuming the existence of a decision space, defined over multidimensional utilities, with alternatives defined in a space with determinable utility values. Ignore, for the moment, the problem of defining the space and measuring the utilities, and focus on the number of alternatives. Most of the poor households that we and other sociologists, economists, and anthropologists have been studying are not confronting decisions at all: they are being impelled by forces beyond their control like hunger, the threatened loss of shelter, fear, unemployment, crying babies, police repression, domestic violence, and a work day that seems interminable. The decision space is defined by the very *absence* of alternatives, and the most we can say, formally speaking, is that the solution is degenerate. A sufficiency of alternatives simply does not exist to qualify the problem for decision analysis. Therefore, the use of decision theory as practice and as metaphor is not appropriate. No one has successfully constructed a decision space that is adequate to describe the situation of poor urban families, let alone carried out a complex decision analysis that would permit them to call their results a study of strategy,[14] and to use a term metaphorically that implies the existence of alternatives when the nonmetaphorical reality is bereft of them is surely to use the wrong metaphor.

The third problem is the *"survival problem."* To what degree are people really surviving? It is true that large numbers of people survive infancy and childhood despite high mortality rates, but if "survival" is to mean anything it must be culturally defined.

There are two problems with its cultural definition. First, it is not a cultural term. People in urban Mexico do not talk or apparently think about themselves "surviving" when they wake up in the morning to face the day, or when they make love, quietly, at night. They are "bearing up" (*aguantando*), "keeping their act together" (*manteniéndose*), "sticking up for themselves and keeping on" (*rajándose*), or perhaps most common of all "looking after themselves" (*defendiéndose*). They may talk of "respect" (*respeto*) or of fights or of living together harmoniously (*convivencia*), but they rarely talk of "survival." So the term does not have ethnographic justification.

Second, given that "strategy of survival" is an analytic term, it must mean survive as a member of a human community and not just biological survival. "Survival" must imply that people are able to participate fully in the cultural life of their community, according to the way their community defines them, as child or adult, man or woman, and so on. But this is precisely what the people with whom we have lived and among whom we have studied are unable to do or are only able to do with great sacrifice. How can one speak of a household or family surviving when the members cannot afford to marry their children properly, according to the prevailing customs of their community? Are families surviving when the children cannot eat festive foods, court, or play because they have to sew peaks for caps or because they do not have enough energy, since there are not enough beans, and meat has not been in the diet for a week? Is a woman like Chabela, who has to work 18 hours a day, "surviving" in a cultural sense?

Since the terms that we use are being used in a metaphorical, informal sense, connotations are important. As the point of our book, and the many good studies that have preceded it, is to show how people are not surviving, are unable to make decisions, and are even more incapable of exercising sufficient control over their lives to formulate strategies, it at least behooves us to be careful in the use of these terms.

But the fact of the matter is that there is not a term that describes what poor households do to get by in either language. This is hardly surprising, since what they do is very complex and varies a great deal from household to household depending on the situation and the human and material resources that the household has to meet the challenges of getting by and getting on from day to day.

Our solution is an unsatisfactory recognition of the lack of conceptual clarity in the area. We use ordinary terms like "getting by" precisely to avoid the connotations implied by the allusive use of formal terms like "rational" and their connotations in ordinary speech. Sometimes people are wily and foresightful and think of strategies for getting on, but much of the time they are just holding on and defending themselves as best they can. There is no global term that catches all of this complexity and promises formal solution programs.

Production and Reproduction

If households are not the sites of rational decision making, how can we characterize their activities accurately? First, they are productive units: they produce goods and services, and as many people have pointed out in their studies of domestic work and the informal sector, their inputs are absolutely essential for the overall functioning (we would hate to say "efficiency") of the national Mexican economy. It is not just textiles that get produced in the household; goods of many kinds are produced in tiny *tallercitos* (workshops) so long as they do not require heavy machinery for their manufacture. Steel work and carpentry products are especially frequent. All manner of services are produced in households, like upholstery, custom ironwork for houses, carpentry, automobile and truck repair, and even the custom preparation of trucks for the road. The successful survival of the Mexican urban household, attaining what we call the state of "getting by," depends on the income-earning activities, or support activities, of all members of the family.

But it would be a great mistake to focus on production alone. Households are the locus of reproduction as well as production, and reproduction occurs in three senses. First, households reproduce themselves in the biological sense: babies are born into households and families, and households are the place and activity sets where they are turned into social, or at least tractable, members of society. Second, households are the sites of reproduction in the sense that they are where activities take place that sustain life, that reproduce and repair people so that they can participate in production activities inside and outside the house. Third, households are where the society is reproduced in a restricted sense. Households strive, among other things, to reproduce their class position in society. In doing so they collectively reproduce some transformation of the social class structure of society itself.

Biological Reproduction and the Household

To speak of reproduction in the three senses suggested is necessarily to speak of women, their biological and social roles, and their subordination in the household, the family, and the society. Lourdes Benería (1979:222) has argued that "women's role in reproduction lies at the root of their subordination, the extent and nature of their participation in production, and the division of labour by sex. . . . Only biological reproduction is necessarily linked with women's specific reproductive functions. Yet most societies have universally assigned to women two other fundamental aspects of the reproduction of the labour force, namely, child care and the set of activities associated with daily family maintenance." Though her cross-cultural examples are not explicitly taken from Mexico, they certainly could have been. The relationship between reproduction, nurturing, and subordination was seen in the description of the "ideal family" in urban Mexico.

The Household as the Locus of Social Reproduction

The household is also the site where the processes of socialization and inculcation of values and ideology takes place in the context of the maintenance of the household and its constituent kin relations. Rather than discuss these processes in the abstract, a more graphic and concrete feeling for the "process of social reproduction" can be had by thinking about the simple process of getting to and from work in the capital. One can start by stepping on the pink line (or first line) of the Mexico City metro at 7:00 at night, in January, in the center of town.

Going Home and Going to Work

The first time you are a part of the evening rush hour going from downtown to the workers' neighborhoods on the outskirts, the experience is frightening, and you emerge in a state of shock. Packed tightly into the metro train, you reach the end of the line, and the human wave frog-marches its way up the stairs to the level of the street. Vendors, miraculously balanced on handrails are selling combs, wallets, credential cases, candies, and penny novels as the wave flows thickly, massively, and quickly past. By the time the street is reached, night has fallen, and the stampede for the buses begins. The ticket collectors lean out the open doors of the dilapidated, diesel smoke-belching buses as they lurch along on clutches rubbed raw by age and adventuristic driving and shout their destinations in a cry that reminds you of the "cries of aulde London" more than the jarring megalopolis. "Romero, la Romero, la Romero-Chimalhuacán" goes one chant, as if some outsider bus could suddenly appear to steal the custom, and the human wave now parts into purposeful streams and fills the vehicles. Some people run down the line as the buses move forward and dive in through the open door, past the ticket collector, and grab a seat. In fact, often the buses never stop to pick up passengers, but just slow down while everyone jumps on as best they can. The bus fills up. You can hardly move. You test for fullness by bending your knees to see if your neighbor's friction is sufficient to hold you up. Not quite. The traffic is disordered, but more than plentiful. The bus must take a left turn into the main artery, and few automobile drivers are going to look upon this maneuver with charity. With a truly baleful stare for the cars, the bus driver challenges them to stop him, grazing a fender here, threatening mayhem and manhood there, all with a rich gesticulatory repertoire. The heat becomes unbearable, even though it is winter. The driver is up to any challenge, knowing that he has an experienced, understanding, and, most important of all, captive clientele. He picks up speed and then stands upon the brake in preparation for entry into a pothole of mammoth proportions. "*Chin*," mutter not a few passengers, "*pinche bache*," who, however well they may have braced themselves, fall forward into a human heap. Fortunately, there are few small children aboard, for it is the *hora pica*, or rush hour. Forty-five minutes later the ride is over and one is home, and the drained weariness of the passengers is evident. Neighbors mutter greetings to each other in recognition of each other's humanity, so recently and outrageously denied by the simple process of coming home from work. Heads down, exhausted, they walk quickly down the muddied, unpaved streets of their neighborhood, seemingly not noticing the perpetual football games under the pale street lamp, the stick games, tops, and hide-and-seek which make up the street life of the young children. The "World Cup" momentarily disengages to let the adults by, and one by one they find doors in the high walls, or *bardas*, that front all the houses and disappear into the one haven they have in this world.

The morning scene is much more orderly, since people go to work in three waves. The first wave goes out between 4:45 and 5:30, catching the first fleet of buses. Those were the ones whose unmuffled crank and starting roar woke the whole neighborhood, including some otiose roosters who find themselves daily made redundant by this technology. These are workers that work in factories in distant places like Tlalnepantla or Azcapotzalco and have to be there by 7:30 in the morning. The second wave of workers goes out around 6:30 to 7:00, for though their work is in the Federal District, it is closer. The third wave, which for the first time includes a number of well-dressed women, goes out at 8:30. These are the bureaucrats and the teachers and the white-collar workers. They pile into the same buses, or walk briskly to the main avenue, which is half a kilometer away and flag a crowded peso-taxi to catch the crowds at the metro station.

The contrast between the broken, exhausted, robotlike human waves of the evening and the brisk, orderly, dignified morning exodus is easily noticed. What is less easily seen is the work of restoration that took place behind closed doors as each householder disappeared into the outside wall, or *barda*, that frames his house lot. The work of restoration that alone makes possible the task of going to work each morning is a fundamental aspect of the process of social reproduction, albeit unrecognized and unpaid.

The worker was probably a head of household (73% of the urban labor force is) and probably male (78% of the urban labor force), and if not a head of household, an unmarried son or daughter. When the householder walked in off the street into his house, he probably owned the house (only 23% of our sample rent), and it was probably what is called a *casa definitiva* (73% of them are), that is, a house that stands on its own in its small lot, or *solar*. In Mexico City it was a little smaller than in the provinces with an average 106M^2 and 2.2 rooms (compared to 146M^2 and 3.1 rooms in the provinces). The house would serve as a workshop for construction, or perhaps as the site of a small business, with the street an extension for enterprises like a small body shop, or a lumber scrap yard, or there might even be as many as four sewing machines in the front room closest to the street where children and relatives (rarely unrelated employees) would be working on a contract to a shirtmaker in the Federal District (6% in our sample indicated that they used their house for commercial purposes). Outside the *barda* is the street, the realm of contestation and disorder that contrasts so clearly with the order and love that dominates inside the wall. No one can fail to notice the difference. Here in Ciudad Netzahualcoyotl, the street, so full of mud, dust, garbage, and excrement, contrasts starkly with the clean, neat house where the pots and pans are lined up in geometric order on the kitchen wall and where the floors are carefully washed every day. The street is the public way and deserves contempt and ordure, for it has already been soiled and polluted by the conduct of the public authorities and the instrumentalities of the state with their sorry record of chronic deceit, corruption, and blatant immorality. The streets are the symbols of the enemy force, almost a force of occupation. They are now laid out in a grid pattern, but it was not always this way.

Originally when the people came here either to escape the rents of city landlords or to build their first secure house in a city where there were greater job opportunities than in the rural areas where they were born, they built their house on lots where they could. Houses grew higgledy-piggledy, not according to some city planner's design, but according to the needs of the family as it expanded with relatives and children. Roads meandered, turning into paths and then broadening again where they must, attentive to the needs of people, not authorities, unplanned and spontaneous. But authorities do not like communities where roads meander and houses become inaccessible and defensible, and where the police cannot enter. So a regularization commission was created to "straighten things out." Titles were granted to lots and houses only on the condition that the roads were straightened out and police access was guaranteed. No rookeries here, no projects where children can flout the law and disappear.

But that turns the private path into the public way, and what was once the way to a neighbor's house or the way home becomes alien territory. Now it is the symbol of public authority, and until such time as the public authority earns the respect of ordinary people, it is treated according to its merits. Garbage is dumped upon it.

The house behind the house walls is order, calm, love, and cleanliness. The world beyond the wall is disorder, disquiet, selfishness, and dirt.

So it is that after an hour or more the householder's face relaxes, and he turns to enjoy the company of his wife and children, for the house is the domain of his authority, and the women and children attend him and assuage his bruised dignity. Liliana Valenzuela (1984) has described one such scene which imparts the flavor of coming home after the subway ride. Ciudad Netzahualcoyotl, where the scene was repeatedly observed, is a 45-minute ride on the third-class bus from the Zaragosa station at the end of the metro's pink line, 45 minutes, that is unless the bus breaks down or unless the traffic is particularly heavy.

"Just look at you, you're soaking wet," exclaims Doña María to her eldest daughter, Concha, who has returned from work early this Friday evening. Concha, still out of breath after the race from the bus stop to the door, sits on the edge of the bed and takes off her muddy shoes, her face framed by wet locks of hair. Doña María kisses her daughter and says "Ay, hija—I'll reheat dinner for you. There's prickly pear soup and tomato stew." Concha puts on dry clothes and sits at the table. Bookshelves, several chairs, a stove and two bunk beds surround her; the kitchen is also the only bedroom. A single lightbulb hangs from the ceiling, casting a diffuse light onto the green walls. Aunt Blanca stirs the *atole*, a soothing drink made from corn flour and cinnamon. One by one the twelve members of the family arrive from work or school, each embraced by Doña María and Aunt Blanca as they enter.

Outside, night falls with the rain and the unpaved streets become rivers of mud—an almost permanent condition in the rainy season. Heaps of garbage mix the mud and dog shit, luring flies and mosquitos. The danger of contagion upsets the mothers of the neighborhood, especially since they pay fees for paving the roads and receive only promises in return. . . .

Doña María's son, Juan, comes in carrying a small bag of groceries. "I couldn't get any sugar at the CONASUPO (the government stores where subsidized and other basic products are sold), and I hear rumors they're hiding it," he tells his father. . . .

The family's schedule is very complicated, but despite that, they gather each night to share a common meal. Here in the home they can be honest about what they feel. Clothes are mended, bellies filled and minds eased.

Later in the evening, Concha and Benjamin leave the table and go into the room in the back to study. This extra room is a welcome addition to their home; once there was only one room and the children could not stay up late to study.

Sleep slowly overcomes those who stay in the kitchen, and it is transformed into a bedroom; they move the table aside, and lay out carpets; blankets and straw mats miraculously appear to accommodate the young kids, while the older ones and adults occupy bunk beds. Parents and other adult couples learn to disguise their lovemaking from themselves and from children in the room. Not many secrets can be kept within these walls.   (Valenzuela, 1984)

But the household is not only some restorative asylum for the victims of modern industrial organization. It is a domestic, social, even political organization of people where interests sometimes coincide and sometimes conflict, where ideology valorizes different sacrifices differently, giving preference to elders and to males in recognition of the importance of the role as producers to the very industrial labor force to which the majority of Mexican men do not belong.

Women work, and not just cleaning and preparing food, shopping and managing budgets, caring for children, and exercising ingenuity in making their households tolerable, even joyful places to love and recuperate. They earn incomes as well, though they are often not on official record as work force participants. They work both inside and outside the house.[15]

Family employment in the small shops (*misceláneas* or *tienditas*) is normally not counted. Nor are the income-generating tasks like running a sandwich stand and preparing food at home for sale. Coverage of activities like shirtmaking and tailoring on the putting out, or *maquila*, system is scant in official statistics, despite the fact that there were as many as 25,000 independent women engaged in putting out work in Ciudad Netzahualcoyotl alone, some of whom hire relatives and family members, all women, for both occasional and steady jobs (Lorenzen, 1986).

The *doble jornada*, or double day of work for women who earn incomes inside or outside the home and supervise or actually carry out all the tasks of keeping the family together, is well known in Mexico. Women work at least as hard as men in Mexico. As Sylvia Chant (1984a, 1984b) has pointed out, auto-construction may be a housing solution for poor people who can build their bamboo shacks and wait until their circumstances improve until they finish the house, but the women pay the price. *Jacalitos* are impossible to keep clean and neat, and although the room is small, the housework is interminable, depressing and never ending.

Men in the work force suffer as well, for they have to endure the outrages of capricious authority that are inevitably fostered by the easy availability of labor, with national unemployment to reach over 19 percent and underemployment 38.5 percent by the end of 1988, according to the Economic Intelligence Unit's Country Report for Mexico.[16] Federal labor law may be liberal, comprehensive, and enlightened, insisting, for example, that in cases where there is ambiguity, all decisions rendered must favor the worker rather than the employer (Schlagheck, 1980). But the liberal and socialist intentions of the writers of the Constitution of 1917 are overwhelmed by the realities of Mexican industrial life in the 1980s, and few workers enjoy the kind of union protection and job security that was envisioned in that most remarkable document. But women, whether they are in the work force or not, carry the burdens of the system, in the sense that it is by their emotional strength, psychological endurance, and physical and moral strength that the workers will be able to board those buses the next morning. They work hard at low wages so as to furnish the subsidies that an inefficient, wrong-scaled, unbalanced, and dependent economy requires so that it can produce goods that can be sold on world markets.

Machismo, the ideology of manliness, plays an important part in this daily dramatic process, and so does *abnegación*, the doctrine that states that a good woman will give up everything for her family. It is useful to think of machismo in two forms, the exaggerated swaggering bully form (the *pendenciero fantarrón*) and the quieter, more reserved form that borders closely on being *formal*, being a serious, mature man whose speech is measured and whose word is secure. A macho of the first sort can and does brutalize his wife and tyrannize his children, is given to drink and boastful lies about his importance in the world outside the home. His reward lies only in the pleasure he derives from his self-indulgence and in seeing fear in the eyes of those he loves. He knows that his wife will find it difficult to escape, particularly when the children are small. She leads a captive life. He often prefers their single-incomed poverty to the presumed assault on his dignity that would be involved by her going into the work force and gaining some freedom from the prison of the household. But this macho, the bully, is doomed, as our analysis in this book shows, for his children will leave him and set themselves up independently and severely reduce his chances for escaping immiserated poverty.

The other kind of machismo, a reserved, authoritative, and confident maleness that eschews drink and brutality and is loving to children, considers their needs, applauds their efforts, and rarely chides them after they come early of age, has its reward too in the loyalty of children and the creation of a multiincome-generating collective where money is pooled and expenses pared so that the whole household can rise out of miserable conditions and be said to be "getting by." Of such families it is said, "They know how to keep themselves going," and the way they live is the answer to the question that seemingly baffles the middle and upper classes, "How do they get along?"

Maleness has its privileges even in the household just described. Although some boys do learn to cook and wash clothes, they are special boys who "understand," and they are doing special things. Men look after children, but the children are not ultimately their responsibility. When the children get sick, men and women both know that, even in the most enlightened households, the women are guilty or at fault in some way and will take the most responsibility. When it comes to deciding about having children, the men's wishes will be more respected than the woman's, at least openly, and the male's wishes for endorsement of his masculinity by the plain evidence of his wife's fertility will be respected, even if this attitude is dying in contemporary, urban Mexico. Men are catered to in ways that reflect an ordering of work that was rural or imported-industrial, where the home was the woman's domain and the workplace was male, and the production of the male was uniquely essential for the survival of the household. The ideologies of the former society survive into the present, and men take advantage of it, some more than others.

The Household and the Labor Force

The household is not just a refuge from the degrading conditions of work in an economy plagued with underemployment and the consequent moral and material exploitation of the workers. It is an arena of conflict and negotiation. But it is not just conflicts between individual members that are negotiated; institutional conflicts meet and are resolved (or partially so) in the household. As Marianne Schmink (1984:87) states,

> The study of household behavior is pursued primarily as a means
> of bridging the gap between social and individual levels of
> analysis. The key concept in making link is that of mediation.
> In response to the opportunities and constraints defined by
> broad historical and structural processes, the domestic unit is
> conceived of as mediating a varied set of behaviors (for example,
> labor force participation, consumption patterns, and migration)
> that are themselves conditioned by the particular makeup of
> this most basic economic entity. The focus on domestic unit
> mediation of individual decisions and behaviors permits the
> study of differential responses to general structural conditions

as well as the analysis of changes specific to subgroups of the population.

The household mediates the demands of individuals, their aspirations, hopes, and goals and the demands, constraints, and opportunities provided by Mexican society. It is in the household that collective decisions are taken whether to satisfy the demand for labor, population, higher or lower skill levels or levels of education, and consumption. People will stay poor, rather than commit the mothers of their families to the labor market; they will raise or lower their fertility to satisfy themselves rather than their expectations of future high incomes (through increased fertility) or prosperous futures for their few, well-educated, and well-placed children; they will deny education to their children in order to insert them and exploit them in the work force, and lastly they will forgo consumption of industrial goods in order to avoid encumbering themselves with debts. But most important of all, they will interpose the collectivist ideology of the household and family between the individual and the state.[17]

One battleground concerns the quantity and training of workers for the work force. Under current conditions of economic development and labor demand, it is in the interest of employers and the state to have furnished a cheap, docile labor force that can be adapted to the technological and working conditions of dependent capitalism, at the same time as it can efficiently (i.e., cheaply) be moved in and out of industry in accordance with cyclical changes in demand and be shifted from job to job according to managerial imperatives and not worker preferences. For historical and political reasons, the costs of inefficiencies in a small dependent economy are paid by labor in the form of unfavorable (cheaply furnished and maintained) employment conditions, work processes designed to optimize production efficiencies no matter what the cost to the worker as well as reduced consumption arising from low wages. Though some of these costs have been offset in the past by the governmental subsidization of basic foods and transport, still labor carries the brunt of the inefficiencies of dependent development. These include, in the Mexican case, diseconomies arising from small local markets, with an emphasis on market widening rather than market deepening over time,[18] inefficient capital markets (where a surprisingly high percentage of capital is generated from sources within the firm), inefficient and costly technology (which is subject to interruptions in maintenance and acquisitions because of chronic balance of payment problems), inappropriate technology (which underutilizes abundant resources, like labor, and overutilizes scarce resources, like capital), undereducated and inexperienced work forces (after all, the median age in Mexico is around 16 years), and hidden taxes in the form of pervasive corruption, particularly in the sectors where licensing or the use of licensed imports is required. These distortions must be paid for, and a glance at the income distribution of the country shows who is paying for them: the poor and the workers. And the costs are paid in the form of low wages, reduced consumption, and docility in the workplace.

How does this happen? What are the mechanisms that bring about this state of affairs so desirable to capital and so necessary for dependent development? Downward pressure on wages is exerted constantly by the presence of large labor supplies and underemployment running between 35 and 40 percent. In addition, the costs of reproducing the labor force are laid off on the work force itself or are socialized. To a large degree the costs of housing the workers are met by the workers themselves or by the state. They construct their own houses on small lots that are, in general, not of great value in the urban real estate market. In fact, it is difficult to know whether there is an urban real estate market for worker housing at all. The state subsidizes food costs, at least it did until the intervention of the International Monetary Fund during the present period of crisis and its opposition to food subsidies. Transport is highly subsidized as well, especially in the metropolitan zone where, until mid-1986 a subway ride cost a peso, worth US$.002, the biggest travel bargain in the world. Basic education is free, until the end of the primary grades, after which costs mount quickly in the system. The major expenditures that every household must pay either are subsidized (food, electricity, water, and gas or cooking fuel) or forgone (telephone, health care, insurance, installment payments, automobile, high-quality staple foods, "luxury foods," now including meat, and first-hand clothing). From the point of view of the working household, they cannot afford much that is not subsidized, since their wages are too low.

With the combination of downward pressure on wages and socialization of the costs of reproduction of the work force, combined with a system that lays most of the real costs onto the workers' families themselves, Mexican capitalism achieves sufficient economies to function in a captive market, and sometimes even in a global market.[19]

Material means alone are not sufficient to order a society on these lines, particularly a society as unfairly ordered as contemporary Mexico. Ideological pressures and mystifications must both placate and motivate people and must serve to justify direct and indirect repressive measures that are necessary for efficiency. The methods of direct repression need little discussion, since they are well known. Official violence in the form of pervasive police brutality, the recrudescence of *caciquismo,* and the concomitant use of private force along with official force in the countryside make Mexico a violent place to do business, as Riding (1985) has said. Indirect repression includes the use of the rumor mill to disseminate the information that the government can and will revoke subsidies and privileges (like entry to secondary school) to either individuals or to everyone if people become too overtly restive. In the end this will be backed either by the army or by paramilitary forces that the state pretends that it cannot control (such as the infamous *halcones* who were responsible for many student deaths in 1968). Rarely is direct repression necessary, most often the threat to take away the milk ticket from families with small children, or to reduce the subsidy for basic products will bring about a more cooperative frame of mind on the part of working people. Ordinary Mexicans know perfectly well that *papa gobierno* can reduce their standard of living to well below the poverty line with a stroke of a pen.

The key to the ideological state apparatus, here as elsewhere, is the educational system. If the argument can be made, as it has been by Bowles and Gintis (1976), that U.S. education that is aimed at creating, encouraging, and rewarding those qualities in students that preadapt them for life in the business and governmental bureaucracies, and if further argument can be made, as it has been by Willis (1979), that this same kind of educational system in its British variant is precisely geared both to middle-class goals and to assuring that working-class children will not only take working-class jobs but will want to take them as part and parcel of their identification with a counterculture that alone makes school tolerable, then an analysis of Mexican education will reveal that advancement depends on the development of those characteristics that are required for efficient operation of the political economy. These are acceptance of authority, acceptance of capriciousness in the working of institutions, attachment to the mystique of *lo mexicano,* and acceptance of the paradoxical notion that a class-ridden divided society that operates to preserve privilege is the logical outcome of a social revolution that achieved democratic goals that are permanently on the verge of being attained.

The mass media play an important and growing role in the ideological scheme. Rather than insist upon loyalty to business and industry, the media foment the message that business, industry, and the state are all working together to create progress, a value that all Mexicans can endorse. More important than the propaganda assaults through news programming and direct advertisements for the state are the favorite television programs of the adult majority, the famous soap operas. These are brilliantly conceived (and executed) propaganda devices in that they are concerned with traditional upper-class values such as filial piety, the importance of honor, the sanctity of blood and blood relations, and the sureness of retribution if the moral constants that define these concepts are violated in any way.[20]

Two important messages are conveyed in the soap operas. First, the upper class has a moral basis, and its worries and priorities are no different from that of ordinary folk (who, when they appear in these programs are endowed with highly moral characters and overarching wisdom about practical moral questions). Second, those symbols and meanings that define the family and kinship and, therefore, underlie the household are determinate in the fate of the household. In other words, the important elements of the good life in Mexico lie within the control of the family or household, and the downfall of the household is the responsibility of its members and not of forces outside it. Victims are to blame for their own circumstances.

The impact of these ideological assaults on ordinary people would be small if it were not for the historical process that has created a system under which ordinary people can only get ahead if they satisfy the demands of business and industry.

The Individual, the Household, and the State

We talk about the "interests" of the individuals in the household and the interests of the household as a whole. This is unexceptionable provided that we do not think of them as always and necessarily the same. But there is another actor or perhaps an *eminence grise* in the household, the state, and it has interests too. The state with its ideological apparatus impinges in an important way on the household and its decisions, and many if not most important decisions are taken with the political and economic constraints of the state in mind. An important aspect of the state apparatus is the labor market. As García, Muñoz, and Oliveira (1982 : 8) put it, "The limits and the possibilities of activity on the part of individuals is given by the structure of the labor market which is determined at the macrosocial level. But the impact of its demands upon the individual is not a mechanical process: it is mediated by the fact that the supply of labor is constituted by individuals who belong to households (*hogares*) where they maintain distinct types of relations with the members." So here we have a triad of actors and an arena. The actors are the individual, the household, and the state, and there are two arenas: the household where labor is supplied and the labor market where labor demand is expressed.

In other words, it is useful to think of the household as an arena where three sets of interests are negotiated: those of the individual, those of the household itself, and those of the state. There are many areas where interests diverge and compete with one another. Children have to compete with each other within the household for funds to enable them to forgo full-time work and continue with their education in order to compete for the good jobs. The parents have an economic interest in retaining the children, their domestic services, and a portion of their earnings in the household in order to raise welfare levels of all the members. The Mexican state has an interest in securing for itself an educated work force at the least expense to itself. And the labor market rewards better-educated men and women with higher pay and better jobs. The problem for the household resides in the fact that, until 1982, the good jobs that compensated higher education levels were disproportionately to be found in the capital city, and even when found they would require that the children move out of their parents' households to take the jobs. Obviously, often the children would want to in the name of social mobility. Parents, in this example, square off against the kids and the state, in the battle for their children (and not least of all a share of their children's earnings). And the battle is not fought in the labor market so much as in the household itself, where children are urged to forgo the potential rewards that the state can offer, forgo as well their own individual self-interests in favor of their family, their younger brothers and sisters, and aging parents. In this instance the household is the enemy of the state, an enemy in the sense that its integrity and continuity is premised on its ability to hold the loyalty of its members and to maintain its own membership. It has to resist the state's propaganda that comes in the form of blandishments to its members in order to keep its own house in order. At the same time it has to operate in a field defined for it by the state, and in particular it has to face up to the hardscrabble world of low-paid employment, underemployment, and unemployment.

The urban household is the arena for these struggles. It is an important struggle, and like so many social and political struggles, it is important that no one wins completely. If the state were to win completely, the household would dissolve and with it the primary bonds of kinship as individuals pursued their own self-interests in a kin-free society. If the household were to win, the state would be the loser, for the household would expend its resources carefully (and probably not invest in the education of the children) and would fail to furnish the trained and socialized workers at the requisite levels of skill needed in the economy. So a bargain is struck between the state and the household such that the state has an interest in keeping the economic, social, and cultural level of the household down so that it will not have to furnish rewards that it cannot afford, but sufficiently high to coax out the necessary labor supply in the desired socialized and educated condition for the labor market.

The bargain that Mexico's ordinary households have made is not a pleasant one. Households are poor, and some are brutalized. The income distribution in Mexico is very unequal. World Bank figures show the ratio of income between that earned by the top 20 percent and that earned by the bottom 20 percent, and it is immediately clear that Brazil and Mexico have the most unequal distribution of income in Latin America (along with Peru and Panama), with the top quintile of the income distribution earning 20 or more times the amount of the bottom quintile, compared to 11 times in the United States and six times in the United Kingdom. Moreover, the distribution of income has deteriorated sharply since 1963; the share going to the bottom quintile having declined by about one-third. In addition, the purchasing power of incomes has dropped seriously in Mexico since 1982, with economists estimating a loss of 40 percent from 1982 to 1987 in individual incomes and our 1987 figures from Oaxaca showing a 23 percent decline for household income, 1977–1987.

As a result of an economy and redistributive system in the throes of development, most observers would agree with Riding (1985:221) that "the middle classes, the professionals, well-placed bureaucrats, middle level company executives, owners of small businesses . . . belong to the wealthiest 30% that earn 73% of the country's total income. Thus, while Mexico's very rich live in a style that would put all but a few American millionaires to shame, and the middle classes enjoy the standards of suburban Americans, its majority lives in degrees of poverty ranging from mere survival to outright misery." Development has favored the richer elements of the country, and the poorer elements have had to suffer the consequences of an unbalanced and unfair economic system.

These are the ground rules of the game for the household. What it does is put up the good fight within these ground rules. And it does this by organizing itself. This response is shown in detail in the rest of this book. In this chapter we have tried to set the scene for the struggle in a theoretically relevant way. In the next chapter we sketch out the instruments for the struggle: the households themselves.

4. The Mexican Urban Household

In this chapter we sketch out the attributes of the Mexican urban household. The two-stage quota sampling strategy outlined in chapter 1, in the cities described in chapter 2, yielded a sample of 9,458 households with sufficient data for analysis. In this chapter we try briefly to describe the household in general, and then we break the concept down into four constituent types: singleton households, matrifocal households, nuclear households, and complex households. Three-quarters of the households are nuclear, made up of a married couple and their children, so it is the typical household form.[1]

The House

The house in urban Mexico, no less than in rural Mexico and elsewhere in the world, is steeped in symbolic meanings (see El Guindi and Selby, 1975, for discussion of rural Zapotec [south Mexican] symbolism). The Spanish word for "house" is, of course, *casa*, and for "household" is *hogar*. Sometimes, people of the house are referred to as *caseros*, or more commonly as *hogareños*, with the implication that the members are related by kinship. In the Zapotec areas, *caseros* is a Spanish gloss for a local term that means both "people of the house" and "people who are in the proper status to participate as insiders in important rituals such as baptism, raising the child, and marriage" (El Guindi, 1977, 1986). Some of that sacredness carries over in the city, just as many urban households, particularly in the provinces, have household altars with the familiar symbols of the domestic cult, most particularly the Virgin of Guadalupe, adorning them and protecting the household. The ritual features of the house have ebbed, particularly in the metropolitan zone, and the notion of household (*casa*, *caseros*) has been replaced by the notion of "family."

The household consists of all the people who are said to "live" there. Usually this is quite clear: a man, his wife, and their dependent children. Sometimes it is not so clear. A godchild may have come to stay, and it is not clear (except to her, perhaps) that the stay is going to be permanent enough to consider her a member of the household. Servants pose problems. They are not all that frequent among the ordinary classes—about 5 percent of the households have them; but they are not called servants, nor are they thought of as such, since they are "visiting" children, godchildren, compadres, or former neighbors in the ranch or city. Their duties are about the same as those of servants in a middle-class household (washing, cooking, child tending, cleaning). They are paid in kind if adults or given occasional gifts if school-age.

How the house was acquired and built is difficult to summarize, since there are as many ways as ingenuity and social circumstance permit. As was discussed in chapter 1, the housesite might have been acquired in an invasion and held until the authorities decided whether to turf out the invaders with force. More often lots are purchased from *fraccionistas* (land speculators), a process that can be complicated by the multiplicity of vendors and purchasers and the dubious legality of many of the transactions. Sometimes a lot will be sold to a number of different householders, a practice that can cause friction and outbreaks of violence. Other times the *fraccionistas* will not have the right to sell the land in the first place, and its former rightful owner will claim fraud. Still other times the ownership of the plot, or indeed the whole area, will be dubious or arguable under the law, and deals of all kinds can be made. The complexity of the whole process can be seen in Ciudad Netzahualcoyotl, where all of the processes mentioned occurred in the years 1960–1970 and led to such confusion and confrontation that the city acquired a reputation for being a "frontier town" of the most violent sort. (See Vélez-Ibáñez [1983] for an engrossing account of life on this urban frontier in those years.) Having secured the lot and borrowed the money to pay for it, people construct a house bit by bit. For the poorest a hut made of waste materials (*desperdicios*) surrounded by a wire fence will do, but not for long. First, a low wall at the front, then a surrounding wall, or *barda*, will be built (if the lot and one's neighbors' lots have been measured). Next, a concrete-floored, brick one-room house (living-kitchen) with an outdoor privy–wash place will be constructed and furnished with bunk beds if there are many children, a butane stove, a TV set, and tables and chairs for eating and studying. This construction of the brick house usually takes place within a year, or faster if the householder has enough money to pay for the materials. It is urgent to be able to "defend yourself," and self-defense requires that you have a secure house that will allow you to leave it in fair confidence that it will not be broken into. This is the reason for the hurry. Building is continuous, with specialists called in for detail work, special carpentry, or special plumbing work when necessary, if they can be afforded. When the children grow to adolescence, the beginnings of an apartment at the back of the lot might be made, and it can be expanded into a two-apartment two-up and two-down house if a "bachelors' hut" and an apartment for a newly married couple is needed. Though the lots are tiny, rarely exceeding 200 square meters, their space is efficiently exploited, even though the average floor area of the house is under 100M², and the median is 64. Over 86 percent of the houses are one story, and 61 percent of them have three rooms or less.

The Family

Familia has a multitude of meanings of which three can be distinguished: *mi familia* means "my spouse"; *la familia* means all the related people who are living together in a household; and *familia* in an extended sense means all those who once lived there in the household but who have moved away, still retaining rights and obligations in the household. In connection with the last, extended meanings, when we asked about emigrants to the United States, we had to phrase the question delicately so as not to offend politically, and we decided on "Are there *familiares* who are living and working abroad?" as an inoffensive query that made sense. It also required that the head of household or the spouse answering the question decided who was or was not a member in good standing of the *familia*.

Households and Their Members

Most households in urban Mexico are based on the nuclear family (74%) and live in privately owned detached houses, or *casas definitivas*, rather than rented quarters (73%). Only in the Federal District is the percentage of apartment dwellers and renters at all high, at 43 percent, compared to 8 percent for the cities as a whole.

Who are counted as members of a household depends on two interrelated factors: the closeness of the interrelated families and the degree to which they share a common budget. Obviously these are related in the negative sense. Where there is no closeness (*conviven-*

cia), there is no budget sharing unless it is hidden, such as when a mother slips some money to her daughter without the consent and knowledge of a jealous stepfather.

A household can be said to be an extended household if there is both *convivencia* and budget sharing among two nuclear families, the junior linked to the senior by a parent-child relation. Arrangements for budget sharing are very complex and vary from household to household. Judging from our 1987 interviews in Oaxaca, about 75 percent of the households have a wife-manager whose responsibility it is to run the domestic economy. One-third of the time, the husband will give the wife a fixed sum of money (*el chivo*) for household staples (*la despensa*) and will dole out money for other expenses like children's clothes and education, fiesta expenses, furniture, and travel. Another third of the time, the husband will give the manager-wife the whole or most of his pay and receive back a portion of it for his own use. Men's expenses are a recognized item in the budget, but a surprising number of men, to show how much they trust their wives, will state that they take nothing back, and further state that they have no expenses. This is rarely true in the metropolitan zone, where transport expenses at least might be covered. We guess that the remaining third of the time other budgetary customs will be followed, including one where the wife-manager has to ask for money to cover all expenses and justify them. Unless they see the husband's pay stub, or control all the money, wives frequently do not know how much the husband earns. In the case that the wife-manager has outside income, it is usually counted in the family resources but discounted to give her some discretionary spending power. There is no notion of "egg money" that we could discover in urban Mexico. Wives complain when the husband expects them to use all their money for household expenses, unless they are pooling incomes completely, and they especially complain if a husband in the work force does not contribute to the *despensa*. Wives also report that when they are discussing a purchase that is not absolutely necessary, it is considered unfair of the husband to ask the wife to make contributions from her *chivo*; money for food and staples is not considered discretionary income. Middle-class people regard income pooling as a desirable class trait, and it may occur more frequently in middle-class households, which are more likely to be middle class precisely because both husband and wife are in the work force and trying to save up "a little capital" which is again diagnostic of middle-class status.

When elements of two nuclear families contribute to the budget, we define the household as "complex." If the household consists of parents, married children and their offspring, and unmarried children, then the household is defined as "lineally" extended, either patrilineally extended if the child is a married son or matrilineally extended if the child is a married daughter.

Extended household arrangements are not easy to sustain, however; tensions between parents and their children-in-law are especially frequent, with mother accusing daughter-in-law of wasting her son's money or of not contributing enough in money and in kind to the joint budget, or *gastos*. Where tasks are carried out together and work is shared, particularly among the younger members of the extended household, where mother-in-law receives her weekly expenses money (also *gastos*) regularly, then there is *convivencia* and budget sharing and one extended household.

For all the possible complexity of urban households, nowhere better seen than in Lomnitz' study (1977) of Cerrada del Condor, in Mexico City, most households are simple, based on the nuclear family. Table 13 shows the distribution of household types based on the kind of underlying family organization.

The household types are defined as follows:

Singletons: Single-person households, that is, households where the respondent reported that he or she was the only person living there at the time of the interview. Sixty percent of the interviewees were men, and 40 percent were women. "True singletons" are households where no emigrants or spouses are recorded as temporarily absent (55% male, 45% female interviewees).

Matrifocal: Mother-child households with no adult male present.

Nuclear: Households of a married couple and their children only.

Complex: Any household more complex than nuclear: patrilineal extended households, matrilineal extended households, households with dependent parents, and "outsider" households which house distantly related kin or nonkin.

Singleton Households

Living alone is almost unthinkable in urban Mexico, at least for ordinary people. It is not just loneliness or the unnatural feeling of living without the social and psychological support of family. This is unbearable, of course, but there are practical reasons as well. Services in Mexican cities are set up with the premise that there will be children (or servants of one kind or another) whose time is valued

Table 13. *Household Types of Urban Mexico*

Household Type	Number	Percentage
Singleton, of which	227	2.4
True singletons	118	1.2
Matrifocal	586	6.2
Nuclear	7,031	74.4
Complex, of which	1,620	17.1
Patrilineal extended	391	4.1
Matrilineal extended	331	3.5
Dependent extended	504	5.3
Outsider households	394	4.2

Note: The stability of the Mexican family has long and often been commented upon. In this context it is interesting to note that the distribution of types in our 10-city sample coincides almost exactly with Lewis' (1952:37) figures for the village of Tepoztlán and for Tepoztecos living in Mexico City:

Household Types for Tepoztlán (1943) and Mexico City (1951)

Household Type	% of all Families Living in Mexico City (N = 64)	% of all Families Living in Tepoztlán (N = 662)
Singletons	0	6.7
Nuclear families	67	70.0
Complex, of which	33	30.5
Lineally extended	13.5	17.2

very cheaply and who will be free to carry out the chores. This assumption is pervasive. Thus, if it is convenient for some service authority to have people queue up for half a day to pay a bill or change their service, children will be assigned to the task. If the light bill has to be paid, one does not use the mails that one could not trust, nor a checking account that one does not have, one sends a child to spend hours at the electricity office in long lines waiting upon the harried and officious clerks. If fresh bread and fresh milk are sold, there is no need for storage facilities on the vendor's premises, they can be sold on a street corner or in the bakery where the children can wait until it arrives and take it home fresh. Many different kinds of errands require children, from the routine gathering of information in a phone-free society ("Go down and see if Maria's father is home from work yet") to long journeys in search of discounted building materials ("Go see if Sr. Octavio is still selling that used plywood"). The community could not function without willing, fetching children, and the children are not unhappy with their lot.[2] To leave the house means to miss the housework and to be able to visit friends or perhaps even promote a furtive courtship. It is nearly impossible to run a singleton household for those reasons.

There are very few single-member households. Our data report 2.4 percent of them, but almost half of these show temporarily absent spouses, or emigrant members who are still considered members of the household, leaving only 116 true singleton households or 1.2 percent of the sample. There are two kinds of these true, singleton households: people who have just arrived in the city (and 67% report that they have lived in the city for a year or less) and older people whose husband or wife has died, divided about half in each category.

Economic and Social Characteristics of Singleton Households

Singleton households are not well off by any standard. Their total income from all sources averages less than a minimum salary, with the head's median income at 0.69 minimum salaries, and the jobs they report are mainly in the less desirable employment sectors, with 50 percent of them working in casual labor or in the low-level service sector, twice as many as in the sample as a whole. They are older, with 44 percent of the heads aged 60 or more, and they are female-headed to a much greater degree than the rest of the sample (40% vs. 11%), as one would expect in a population created in part by the death of a spouse. They are the only household type that shows no outside income source, either from children, windfall, or other relatives. Survey statistics that compare their social and economic characteristics with matrifocal, nuclear, and complex households are found in Table 14. Their objective condition corresponds to the popularly held opinion of their undesirability. Living alone is not a happy solution to the problems of Mexican urban living.

Matrifocal Households

Just over 6 percent of the households in the sample were matrifocal, consisting of a woman and her children with no senior male members. Not only is this a very low figure compared to U.S. household structure, but 44 percent of them are headed by widows, 27 percent by never married, 17 percent by women currently married, and 12 percent by divorced women.

Table 14. *Comparison of Singleton to Other Households*

Variables	Household Characteristics	
	True Singletons (N = 118)	Other Households (N = 9,341)
Income		
Median household income	$1,237 (US$54)	$3,649 (US$161)
Marital status		
Married	0%	86%
Divorced	9%	2%
Widowed	49%	5%
Single	42%	7%
Age of head		
60 and over	44%	9%
Median age	59	40
Sex of head		
Male	60%	89%
Female	40%	11%
Amount of time in city		
1 year or less	67%	73%
Mean years in city	7.9	4.6
Unemployment		
Head not in work force	35%	12%
Occupation		
Casual labor	17%	8%
Service job	33%	17%
Blue-collar	6%	19%
White-collar	30%	39%
Business/professional	10%	10%

The households are a little older (median age of 43 vs. 41 for non-matrifocals), smaller, with fewer children, and they are more mature, having fewer households in the earliest stages of formation and more in the stage where the children are grown up and potential income earners (28% vs. 14% for nonmatrifocals).

The arguments concerning the feminization of poverty that locate the correlates and causes of poverty in the predominance of matrifocal households in the United States (see Moynihan, 1986; Auletta, 1983) do not hold for urban Mexico. Matrifocal households are not worse off because of the absence of a man, despite discrimination against women in the work force and the difficulties that women have getting well-paid employment. Although median household income is 14 percent lower than nonmatrifocal households, since they average one less member in the household, their per capita incomes are 8.2 percent higher. They put almost as many members into the paid work force (1.38 vs. 1.40), and the ratio of dependents to members in the work force is lower than the nonmatrifocals.

Budgets are slightly tighter for matrifocal households. They spend 6 percent more of their household income on food, but their discretionary income is higher. Employment patterns for the heads of matrifocal households are what one would expect for women in the Mexican economy.[3] More have service jobs, and fewer have blue-collar jobs. Employment rates for heads are quite low at 68 percent, but this is not surprising, since twice as many of them are 60 or older as for nonmatrifocals. Their lower employment rates are offset by higher employment rates of children and other members of the household. Though there are slightly fewer members in the work force, still fewer children (and fewer dependents) in matrifocal households ensure that their dependency ratios are 25 percent lower than nonmatrifocals.

The genealogical complexity of matrifocal households is quite outstanding, with more grandchildren, more siblings, and more outsiders as members, and even if their median household incomes are only 88 percent of the nonmatrifocal households, their per capita incomes are 8 percent higher. Though they are smaller, they are not abandoned and alone and preyed upon by society. Most of them are viable, operating units, doing as well as nonmatrifocal households in the very tough world of the "popular classes" of urban Mexico.

We had thought that they would be highly vulnerable and living on the margin of existence. We had been persuaded of this notion by interviews with people from nuclear or extended families, and case studies, carried out mostly in Oaxaca, of families in tragic circumstances, recently abandoned by a drunken and abusive male, cast upon their own resources, or eking out a pitiable existence taking in washing and making tortillas to sell to other households. To our great relief it seems that such circumstances are relatively transitory, particularly if people are long-enough established to have kin-

Table 15. *Statistics on Matrifocal Households*

Variables	Matrifocal	Nonmatrifocal
Median income household	$2,236	$2,651
	(US$102)	(US$117)
Median per capita income	$541	$500
	(US$23)	(US$22)
Budget data		
Median expenses	$1,494	$1,719
	(US$66)	(US$76)
Median expenses as percentage of		
median income	64%	65%
Median percentage income spent		
on food	47%	41%
Employment		
Mean number in work force	1.38	1.4
Job classification (head)		
Agricultural	3%	5%
Casual	8%	8%
Service	29%	18%
Blue-collar	6%	20%
White-collar	44%	40%
Business/professional	10%	11.5%
Heads not in work force	32%	11%
Household demography		
Mean number members	4.3	5.3
Mean number dependents	2.9	3.7
Dependency ratio	2.0	2.7
Marital status of head		
Married	18%	90%
Single	27%	6%
Divorced, separated, widowed	55%	4%
Median age of head	43	41
Genealogical complexity		
Siblings present	23%	7%
Grandchildren present	10%	4%
"Outsiders" present[a]	23%	7%
Domestic cycle		
All children over 15	31%	14%

[a] "Outsiders" are collateral kin, affines, fictive relatives, or unrelated persons. They and their contribution to the household are discussed in the last part of chapter 4.

folk upon whom they can rely. The percentage of households in the most vulnerable stage of the domestic cycle, when they have small children only, is two and a half times smaller among matrifocals than it is among nonmatrifocals (14% vs. 38%).

Chant (1984a, 1984b, 1985) made a special study of matrifocal households in three *colonias populares* in Querétaro. The results for her subsample of 22 female-headed families (defined as our matrifocals) are strikingly similar to our own, and her detailed interviewing makes it possible to extend our more skeletal conclusions. Like us, she concluded that although individual incomes of the women were lower, the economic deficit was more than compensated by the contributions of the children and other members of the household so that the matrifocal households in Querétaro were better off in per capita terms, just as we found for the city sample. She further discovered that fully one-third of the matrifocal households were formed at the initiative of the women and that "despite major structural constraints of the economic and social potential of matrifocal families, single parents units often fare better than male headed nuclear households" (Chant, 1985 : 650). Table 15 gives comparative data from our study.

Older Matrifocal Households

Because the Mexican family depends so much on its children for support in its old age, and because the process of urbanization and migration has done so much to break up families, particularly extended families in the past 30 years, we were interested to examine the matrifocal families where the head was 60 years of age or older.

What we find is that the same tendencies as were found in matrifocal families in general are found here, but in accented form. The older matrifocal households have fewer children present, but still manage over one per household, and most of these children are in the work force. These older matrifocal households have lower incomes to be sure, but they are smaller, with fewer dependents, and with dependency ratios lower than younger matrifocals. By the usual budgetary measures, they are no worse off than other matrifocals in the proportion of their household incomes spent on food and other unavoidable expenses. The level of genealogical complexity remains high, and 67 percent are widows. In short, they are just like other matrifocal households only more so, smaller, but "more intensely matrifocal." But once again, not abandoned and vulnerable as we had thought beforehand. Table 16 shows the data.

Table 16. *Older and Younger Matrifocal Households*

	Older Matrifocal (N = 105)	Younger Matrifocal (N = 577)
Economic		
Median income of household	$1,995 (US$88)	$2,569 (US$113)
Budget		
Median unavoidable expenses	$1,698 (US$75)	$2,130 (US$94)
Income spent on expenses	85%	83%
Income spent on food	44%	47%
Employment		
Mean number in work force	1.3	1.4
Heads not in work force	70%	24%
Household demography		
Mean number members	3.4	4.4
Mean number dependents	2.2	2.0
Dependency ratio	1.69	2.14
Median age of head	66	40
Genealogical complexity		
Siblings present	5%	27%
Grandchild present	23%	7%
"Outsiders" present	66%	15%

Nuclear Family Households

It is difficult to overemphasize the importance of the nuclear family in Mexico, for it truly is the emotional center of the psychological and social life of all Mexicans. When we talk about the economic importance of complex households and their constituent extended families, we perhaps need to remind ourselves that the extended family is made up of nuclear families and that "completely extended families" in the sense of families that have a single authority structure depending on the *padre de familia* with a single budget administered solely by the older generation into which all earning members of the family put all their wages is very rare indeed. (Aside from those families that are normally found in the higher reaches of the economic structure who own their own businesses and where the *padre de familia* is effectively the chief executive officer of a business corporation.)

The nuclear family is nearly universal in Mexican society. Half the matrifocal households and half the singletons were originally nuclear or are only temporarily in their nonnuclear state. The complex household contains at least one nuclear family. So 96 percent of the households that we studied were either made up of nuclear families or were households that one could reasonably expect would return to nuclear status. The nuclear family is the ideological core of Mexican society, and even gender constructs can be safely regarded as dependent upon and not constitutive of the nuclear family.

The statistics on the nuclear family can be briefly summarized, since they have been prominent in the contrasts which have preceded. Households based on the nuclear family make up about three-quarters of the sample (73.3%), as is seen in Table 17.

Since the nuclear family has been the reference category throughout this discussion, not much further needs to be said, and the statistics are appended to the discussion in the interests of completeness. It can be noted that 98 percent of them have male principal wage earners, as compared to 62 percent of the nonnuclears, and so the bias in thinking of the typical Mexican family as being headed in both the cultural and the economic senses by a male is not misplaced. The median earnings of the head are also higher in nuclear families than they are in the contrast group, but as Chant (1985 : 642) reminds us and interviews in Oaxaca, Mexico City, and Guadalajara confirm, one should not confuse the higher earnings of a head of household with better living conditions for all, since some males are reported to retain as much as 50 percent of their earnings for their own use. Since the number of members actively in the work force is lower for nuclear family households (1.3 vs. 1.6 for the other households), both worker dependency and age dependency ratios are higher for the nuclear group. Median household incomes are actually lower for the nuclear households as compared to the others, which is not surprising since 64 percent of the contrast group is the complex household formation. This also accounts for the lower median household income as well ($3,629 vs. $3,739). With more members per household, per capita incomes are even lower, as would be expected.

The vast majority of married couples have children living in the house (89% vs. 63%), but by definition have no other relatives living at home. Their budget profile is about the same as the contrast group, as are their housing conditions, with no discernible differences in the availability of electricity, water, and sewer services nor with kitchen and bathing facilities.

Table 17. *Nuclear Family Households*

	Nuclear Households (N = 7,031)	Other Households (N = 2,428)
Economic		
Median income of household	$3,629 (US$164)	$3,739 (US$172)
Budget data		
Median unavoidable expenses	$2,211 (US$105)	$2,064 (US$95)
Income spent on expenses	54%	56%
Income spent on food	41.5%	42.1%
Employment		
Mean number in work force	1.32	1.61
Job classification of head		
Agriculture	5%	4%
Casual	8%	8%
Service	16%	21%
Blue-collar	20%	17%
White-collar	40%	40%
Business/professional	10%	11%
Heads not in work force	10%	19%
Household demography		
Mean number members	5.4	5.0
Mean number dependents	3.8	3.2
Dependency ratio	2.9	2.0
Median age household head	39	41
Genealogical complexity		
Children present	89%	63%
Siblings present	0%	19%
Grandchildren present	0%	9%
"Outsiders" present	0%	19%
Domestic cycle		
Child under 15	13%	22%

Complex Households

If you listen to traditional parents talking in optimistic terms about the future of their families, they sometimes talk about the marriages of their children and their children's futures with an eye to keeping the family together and bringing daughters-in-law into the households in one big happy collective. For many Mexicans this is an ideal that is realized in what anthropologists call a stem family, usually consisting of one married son or daughter, their children, and the younger unmarried children. As the older couple's younger children get married, the older married brothers and sisters will move out to establish independent houses, once they have had time to save the money and garner the credit from the parents to afford their own place. A younger married couple takes their place. Immediate independence is valued by some parents for their children; eventual independence is desired by most, provided that there is some child in the parental home.

Retaining the loyalty of children and creating living and psychological conditions for their continuance under the parental roof is not easy. The city affords more economic choice than the rancho ever did, and in particular, it affords more choices that are not under the control of the father. Perhaps for this reason, the patrilineal bias has disappeared from urban households. There are almost as many married daughters living at home as there are married sons, and together, the number of lineally extended households only amount to 7.6 percent of the total (see Table 13, where 4.1% of the sample households are patrilineal extended and 3.5% are matrilineal extended).

There are two ways of forming lineally extended households: the traditional way where the children remain economically dependent on their parents and the modern way in which children's wages exceed those of their parents, and in the latter sense the parents become economically dependent upon their children. In the parent-dependent case, the family is no less an extended family, but the headship of the group has been transferred to the children because a child has become the principal wage earner. There are 504 such families in the sample (5.3%).

Aside from lineally extended households, there are complex households that include an "outsider," that is, a person who is either more distantly related to the head such as an aunt, a cousin, or other collateral relation, or someone who is not related but shares relations of *compadrazgo* (a godchild, a *compadre*), or someone who is no relation to the family at all, a neighbor perhaps, or more likely, a neighbor's child. There is still a tradition of charity among the popular classes, and people who have nowhere to turn are usually taken in, at least in those places where some kind of community has had a chance to form. People look after their own, and "their own" is a term that can be quite widely defined.

We shall deal with outsider households in the next section. Here we continue with a brief description of the characteristics of lineally extended households, which number 1,226, or 12.9 percent of the total sample.

Lineally Extended Households

Lineally extended households are better off than matrifocals and nuclears. Their characteristics are seen in Table 18. Though their heads have no better jobs and no better education and receive no higher pay, lineally extended households are well organized to produce income, with the result that they have 38 percent more members in the work force, and their median incomes are 25 percent higher than the rest of the sample. They have 64 percent more children in the work force which accounts for the entire difference in work force participation, since the rate at which spouses are in the paid work force is no different (at 5%) from nonlineally extended households.

The budgetary profile for lineally extended households shows that they are able to achieve impressive economies of scale in consumption as well. With 10 percent more members, their expenses are only 3 percent higher, and they spend 5 percent less of their household's income on food. So far as housing assets are concerned, the value of their house and lot in constant 1982 pesos exceeds that of the nonlineal households by over 9 percent.

The composition of the lineally extended households is complex, as one would expect. Half have parents present, and 30 percent have siblings, 26 percent have grandparents, and 9 percent have grandchildren. These corresponding figures for households that are not lineally extended show trivial percentages in all these categories. Lineally extended households are twice as likely to have migrants than the others.

Finally, the dependency ratio of nonworkers to workers is much lower in lineally extended households than it is in the contrast set (2.1 vs. 2.8 nonworkers per worker). In short, lineally extended households are able to pool income and cut expenses in a significant fashion and therefore represent an economic, as well as a social and psychological goal, for urban Mexicans of the popular classes.

Outsider Households

Our earlier examination of the role of "outsiders" in households in the city of Oaxaca (Hackenberg, Murphy, and Selby, 1984) led us to believe that they would play an important role in the Mexican urban household. We felt that householders would attempt to recruit out-

Table 18. *Characteristics of Lineally Extended Households*

Variables	Lineally Extended Households (N = 1,226)	Other Households (N = 8,224)
Economic		
Median income household head	$2,897 (US$133)	$3,005 (US$138)
Median household income	$4,522 (US$205)	$3,630 (US$169)
With less than 1 min. salary	19%	25%
With over 4.8 min. salaries	25%	14%
Budget		
Median unavoidable expenses	$2,422 (US$107)	$2,346 (US$108)
Income spent on expenses	64%	72%
Income spent on food	37%	42%
Employment		
Number members in work force	1.8	1.3
Number members not in work force	3.75	3.65
Number children in work force	0.36	0.22
Member under 21 in work force	26%	13%
Household demography		
Marital status of head		
Married	67%	88%
Widowed/divorced	6%	7%
Single	27%	5%
Number household members	5.7	5.2
Households with migrants	23%	12%
Genealogical complexity		
Parents present	50%	1%
Siblings present	30%	2%
Grandparents present	26%	2%
Grandchildren present	9%	1%

siders to the household either to add to the income-generating ability of the household, or to take over housekeeping tasks so as to free members of the household for paid work. There appears to be some truth to this. "Outsiders" appear in households where the income of the household head is 10 percent lower compared to households without them (see Table 19). Although only 36 percent of the out-

Table 19. *Households with Distant Kin or Nonkin ("Outsiders")*

Variables	"Outsider" Households (N = 394)	Other Households (N = 8,556)
Economic		
Median income household head	$2,798	$3,118
	(US$124)	(US$137)
Median household income	$4,299	$3,628
	(US$186)	(US$160)
Employment		
Mean numbers in work force	1.9	1.4
Wife in work force	12%	6%
Member under 21 in work force	34%	13%
Child in work force	27%	15%
"Outsiders" in work force	36%	N/A

siders are in the work force, the incomes of the households where they live are 18 percent higher than other households in the work force. By all means they are not all young people; their median age is 22, while the median age for the whole population is around 16 years. Only 10 percent of the outsiders are under 13 years of age, and 9 percent are over 60, so one cannot say that they are normally dependent on the household they join. The 36 percent who are in the work force average $4,357 per month, which is 3 percent more than the average for the sample as a whole, and this makes their contribution to household income as substantial as family members. Around half the outsiders are married, but very few indeed have their spouses living in the same household, since only 14 percent of the outsider households have more than one outsider.

Household Demography of "Outsider" Households

The labor force statistics on the "outsider" households can be summarized briefly. They have 36 percent more members in the work force, twice as many wives in the work force, almost three times as many young people (under 21) in the work force, and twice as many sons or daughters. Their dependency ratios are 20 percent lower than other households, even though they have 15 percent more members.

The "outsider households" show a good deal more complexity than the others as well, with 87 percent more unmarried heads, more than twice as many female heads, and almost three times as

Table 20. *"Outsider" Households: Composition and Budget*

	"Outsider" Households (N = 394)	*Other Households (N = 8,556)*
Household demography		
Mean number members	6.0	5.2
Mean number not in work force	3.9	3.7
Dependency ratio	2.1	2.6
Marital status of head		
Married	68%	86%
Single	13%	7%
Median age of head	47	41
Household female headed	24%	11%
Matrifocal households[a]	19%	7%
Genealogical complexity		
Spouse present	55%	81%
Parent(s) present	12%	7%
Son/daughter present	70%	83%
Sibling(s) present	12%	5%
Grandparents present	7%	5%
Grandchildren present	16%	2%
Domestic cycle		
Couples w/o children	16%	10%
Dependent children only	38%	69%
All children under 15	32%	14%
Budget		
Median unavoidable expenses	$2,515 (US$116)	$2,348 (US$108)
Income spent on expenses	59%	65%
Income spent on food	43%	42%

[a] A female-headed household is one in which a woman earns the most money. A matrifocal household is one in which there is an older woman living with her children with no husband or other male reported present.

many matrifocal forms as the other households. And as Table 20 shows, more of them have siblings, parents, grandparents, or grandchildren present than the others as well.

Domestic Cycle and Budget

Perhaps the most important thing about these outsider households is the stage of the domestic cycle to which they belong. There are

55 percent fewer households with children in the dependent years until they are 15, and there are over twice as many households in the most productive years, when the children are all over the age of 15. As a result their household economies show more flexibility: though their household expenses are 7 percent higher than other households, the proportion of household income that is spent is 6 percent less.

Conclusion

The purpose of this chapter has been to introduce the Mexican urban household and describe its composition, labor force characteristics, and the bare lineaments of its domestic economy. Appendix 3 presents as complete statistics as this project can on the household in general, as of 1977–1979. We close by emphasizing two things. First, however ingenious Mexican householders are in looking after their own and in ensuring that everyone of the have-nots in Mexican society can at least have the benefit of living in a family-based household, there can be little solace in the fact that they have to. Second, as we show in chapter 9, the importance of the household and family has been in no way diminished by the economic depression that began in 1982. In this chapter we have seen the Mexican urban household at the end of 40 years of unparalleled economic growth from 1940 to 1982. In chapter 9 we shall see that the same rationale for survival, in an exaggerated form, has persisted into the 1980s. The severe tests imposed by the economic crisis have increased the need to form complex households of many members, even as the urban birth rate is falling fast. The logic of the "Mexican solution" among the popular classes has led to a preference for larger households made up of smaller families—an increase in complexity along with a decrease in fertility.

5. Household Dynamics and Economic Survival

In this chapter we show that larger households are better off in urban Mexico because children do not occasion the kind of expenses among ordinary families that they do in the middle and upper sectors. We then show that households improve their economic prospects with age, as we examine the domestic cycle effects on the household's economic condition and structure. We try to be careful not to draw inferences about changes over time from cross-sectional data, but do not always succeed.

The key element to improving the economic condition of the household is time and keeping the household together in a cooperating collective, achieving economies of scale in consumption, and a degree of income pooling that makes everybody better off. Obviously only some households can do this. When the children are young, they cannot earn income. But by the time they become 12 years of age, they are able to get some paid work and are able to contribute not just to their own costs but also to the general costs of the household. Children of that age can do errands, asking people who do not have young children whether there are errands that need doing, as there usually are, since Mexican society is premised on the availability of children to do them. Boys can shine shoes, sell papers, and help their fathers and relatives, while young girls can sell fruit, gelatin deserts, watch children, and sweep patios as well as numerous other tasks. Children are not paid a wage for these jobs, they receive a few cents to half a dollar for their work, and this money is expected to go toward defraying some of their expenses, even if they do not actually turn it over to their parents.

As the children get older, it becomes more difficult for the senior household members to keep them in the household. The children will be expected to work and help out. But this is a deceptive figure. Practically no child of an ordinary Mexican household does not contribute materially to the household in a systematic way, and most often it will be through helping parents or other close relatives. It is almost unheard of for children to refuse to help; it is not exaggerating to say that such a child would be regarded as seriously defective were he or she to do so. In fact, some of the most difficult arguments in ordinary households arise out of the desire of the child to drop out of school to help his parents. "He is my right arm," as one father, who was beginning to do well in the shoe repair business, said about his 15-year-old son, "and I don't know what I would do without him, and he does not want to go to secondary school, but I worry because I know how hard it is to get a job that is not a simple, unskilled one, without a secondary education." (As this father found out when he was laid off in 1985 and with a third-grade education was unable to find work like that which he had earlier had and was forced to start his own small business repairing footwear.) All children who are going to school past secondary are in the labor force, many of them full time. Indeed, not being in the labor force at that age is a defining characteristic of middle-class status. (Postsecondary education is *educación superior* and is roughly equivalent to U.S. senior high school or fourth and fifth form in Great Britain.)

Mexican households exploit themselves to get by, and exploiting one's children is part of the idea. Smart parents here as elsewhere are careful that they not do it with too heavy a hand. Children who feel a lot of pressure from their parents resent it and will eventually leave home. Just as the bullying macho has his comeuppance when his wife leaves him to form a matrifocal household, so too the parents that try to exploit their children excessively and who deny them opportunities arbitrarily will see them depart.

Upward economic mobility, or achieving a minimally decent level of living, is not to be attained through career advancement for the ordinary Mexican household. There are, as will be shown in the next chapter, strong age effects in the individual income equations, but one should not confuse statistical significance with a pay raise because, in fact, the amount of increased income that comes with seniority is quite small. (This is, of course, particularly true in more recent years when employers have been upgrading their work forces by laying off higher-paid workers in favor of younger, better-educated, and lower-paid ones.) Wages in the Mexican economy tend to be tied to the minimum wage schedule, which now differentiates very little between the three minimum wage areas of the country and which stipulates very small pay differentials for higher skills and greater experience. A bulldozer operator, for example, is slated on the schedule to earn about twice what an unskilled operative earns, and even though workers will earn more than the minimum salary when labor is tight or skills in short supply or when employee loyalty is important, the minimum salary has a definite leveling effect on the wage structure.

The only way that one can be sure that one is going to raise one's level of living decisively is through one's household. And the result can be seen if you look at the differences between better off and less well off households: the better off ones are larger, have more children, more members in the work force, lower dependency ratios, and a higher percentage participating in migration. We have partitioned the households into three economic classes, following the criteria of the Fund for Housing (FOVI) for the allocation of credit. They have been allocating credit to households where the total income lies between 1.8 and 4.8 minimum salaries. Their reasoning is that households with less than 1.8 minimum salaries are so poor as to be unable to make any payments and that households with more than 4.8 minimum salaries can borrow from private banking sources. Thus we take 1.8 minimum salaries as the household poverty line and 4.8 minimum salaries as the middle-class threshold. Of the 9,458 households for which we had adequate data on household income, 5,309 (56%) were in the first income class, and we called them "the poor," while 3,245 (34%) were in the second class, and we said they were "getting by," while 904 (10%) were earning more than 4.8 minimum salaries, and we called them "middle and upper class."[1] By every measure the better off households were larger than the less well off ones, and the household heads were less than two years older than in the poor households. Table 21 gives the relevant statistics.

Table 21. *Household Size and Economic Status*

	Income Class			
Variable	*Poor*	*Getting by*	*Middle/ Upper*	*All*
Household size	5.0	5.9	6.4	6.0
Number children present	2.7	3.1	3.1	2.9
Number in work force	1.2	1.7	2.0	1.4
Number dependents	3.9	4.2	4.4	4.0
Dependency ratio	3.1	2.5	2.2	2.9
With migrants	11%	18%	23%	15%

Note: One-way analysis of variance on all variables showed that differences between groups was significant beyond the .0001 level.

The better off households are larger, with more children and more members than the poorer households. The secret is not difficult to find: they have more members in the work force, and their dependency ratios, taken as the ratio of members not in the work force to those in, are lower than the poorer households. The families associated with the households are also larger, as indexed by the fact that they report a higher frequency of members who are working and living in cities of other countries, as the "percentage with migrants" category records. Clearly the better off households and families have more members (but lower dependency ratios).

The argument can be made, and has been made, that stratifying households on the basis of their total income is misleading, for it is per capita income that matters, not total household income. But we would argue that per capita income is a misleading stratifying variable to measure the overall welfare level of a household because it treats all household members equally in figuring the costs of running the household, and this is simply not true in Mexico. In fact, the secret to the success of the larger household in Mexico is that the children add little to the expenses of running a household particularly in ordinary households, or, in our classification, in the households of the poor and those who are getting by. Although we did not obtain cost estimates for each child, we can compare the amount of what we call "household expenses" for households of different sizes. These expenses are those that are met by all households: food, clothing, gas, electricity, and water, and these make up between 60 and 65 percent of the household budget.[2] If one compares these expenses across households with different numbers of children, one finds that they do not rise at all significantly. Table 22 gives the relevant figures for all households.

Even more convincing data can be found in Figure 2, where the changes in the budget figures are broken down by economic class, and one can readily see that for the two majority classes, the poor and those getting by, there is little change in the household budget from the third child to the seventh. The reasons are not mysterious. The fixed costs of running a household do not change with the addition of a child. Housing costs remain the same, since room can always be found for a new child. Most of the other household bills are insensitive to the addition of a child: rent, taxes, electricity, water, gas, and installment payments. Medical bills could be an important additional expense and sometimes are. But the sad fact of the matter is that medical expenses are not an important part of the household budget, since they are forgone by the ordinary families except in

Table 22. *Expenditures on Children*

Number of Children	Household Expenses as % of Income	Expenses	Change %
0	63	$2,566	—
1	60	2,751	+7
2	60	2,892	+5
3	61	3,071	+6
4	65	3,294	+7
5	68	3,229	0
6	66	3,320	+3
7	66	3,361	+2
8	64	3,498	+4

emergencies, and then often subsidized or free medical help is available. Food costs increase, of course, but substitutions are made. The amount of meat does not expand to the number of available mouths, rather it is a fixed amount which gets stretched. Edible items that are bought extra for each member (tortillas, bread rolls, gelatin desserts) are either cheap or subsidized or both.[3] Clothing expenses are important, but by the time the child demands high-quality clothing and new clothes, he or she will be in a position to contribute to its cost if not defray the expense entirely. Education costs are important, of course, but the educational standard for ordinary children is three years of secondary education, and the fees for the cheaper (and most accessible) secondary schools are not high, particularly in the provinces.[4] Elementary education was free at the time of the 1977–1979 survey; the *boletos* had not been introduced yet.

Musgrove's (1978) more detailed analysis agrees broadly with ours. In his breakdown of expenditures, he notes that there are reductions in per capita expenditures as the number of children in the household increases, with economies of scale lowest in food, intermediate in clothing, and highest in housing. When he examines the income elasticities of food, he finds that lower-income people have relatively reduced consumption of meat, eggs, dairy products, and fruit, while the income elasticities are lowest for cereals and pastas. One may infer that people will substitute cereal and pastas for the first food group when income declines, which is precisely what we have found in interviews in 1987 in Oaxaca.

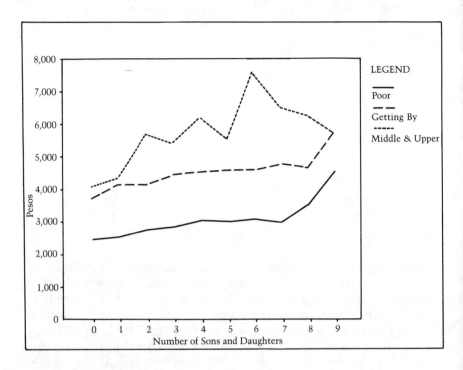

Figure 2 **Expenses by Co-resident Children**

Children in ordinary households in urban Mexico do not cost much, and they are a great potential benefit to the household.[5]

The Domestic Cycle and Improving Household Fortunes

The corollary to this finding that larger households live better is that the household's fortunes improve through time. Table 23 shows the distribution of the 8,306 households for which we have complete data on the domestic cycle.

We partition the households into six decennial cohorts (except for the oldest which is open ended), starting with those where the senior woman of the house is between the ages of 15 and 24 until the senior woman is 65 or older. The economic and social fortunes of the household can be treated under four headings:

1. Increases in household earnings as the household matures
2. Increases in the value of household assets and expenses as the household matures
3. Labor force participation rates of households as the household matures
4. The changing composition of the household as it matures

With very few exceptions, we will find that household fortunes, household size, and household complexity increase with the age of the women, either when the woman enters the 55–64 cohort or, more rarely, in the previous one.

Table 24 gives some figures on incomes. There is a downturn in total household income in the years when the senior woman is between 55 and 64. This downturn continues in her later years, but our figures note that, at the same time, the downturn in per capita in-

Table 23. *Domestic Cycle Frequencies*

Senior Woman's Age	Number of Households	%
15–24	1,313	16
25–34	2,703	33
35–44	1,932	23
45–54	1,352	16
55–64	571	7
65+	435	5

Table 24. *Household Incomes through Domestic Cycle*

Woman's Age	Mean Income Head	Mean Income Household	Mean Income per Capita	Mean Income Discretionary
15–24	$3,998	$8,255	$1,919	$1,056
25–34	4,789	8,436	1,622	853
35–44	4,789	11,023	1,722	916
45–54	4,184	13,452	2,538	1,364
55–64	3,398	9,580	2,456	1,644
65+	2,848	8,117	2,536	784

Table 25. *Household Assets and Expenses*

Woman's Age	Value House and Lot	Expenses	Income Spent on Food %
15–24	$142,764	$2,258	43
25–34	157,269	2,572	58
35–44	165,302	2,814	38
44–54	163,361	2,711	45
55–64	169,118	2,342	42
65+	178,768	2,028	44

come is slight, and there is a continuing increase in the amount of disposable income during the years before she turns 65.

Not only do incomes increase as the household matures, the value of the house and lot increase as well. Table 25 provides the figures on household assets, which are based on the reported value of the house lot and the improvements on the site. These figures are based on what was told us in the survey and should be taken with many reservations, but they do confirm the trend of improvement with maturation of the household. Table 25 also includes data on household expenses, and it is well to note that they peak in the years when the women are between the ages of 35 and 44, except that, as a proportion of the total household income, they too peak in the later years.

As Table 26 shows, labor force participation rates also increase as the household matures, peaking in the years when the woman is be-

Table 26. Household Composition Variables across Domestic Cycle

Woman's Age	Household Size	Number of Dependents	Number of Children	Singletons %	Matrifocal Households %	Nuclear Households %	Complex Households %
15–24	4.3	2.9	1.9	1	0	77	22
25–34	5.2	3.9	3.0	1	3	85	11
35–44	6.4	4.5	4.1	1	9	81	9
45–54	5.3	3.6	3.0	1	13	70	16
55–64	3.9	2.5	1.6	5	18	57	20
65+	3.2	2.2	0.8	10	25	40	25

tween the ages of 45 and 54. In that same age range, the number of coresident children in the work force peaks, and the percentage of households with wives falls only 1 percent from its peak of 10 percent in the previous decennium. It can be noted that the work force participation rates for wives is never very high but that it peaks, as expected, in the years when the household is at its most prosperous point, when the woman is between the ages of 35 and 55.

Table 26 shows some relevant statistics on the size and composition of households as they mature. The dependency structure is crucial to the household's well-being in its middle years. One should note, for example, that the households are at their largest in the years when the woman is between 34 and 44, when the number of children and the number of dependents are at a maximum and the proportion of households that are still in their nuclear phase is very high as well.

We can note also that as the household matures it tends more and more to take a nonnuclear form. The older households, where the women are more than 65 years of age are exemplary: although 40 percent of them keep nuclear households still, 25 percent have reached the desired state of living in a complex household (most of which are extended, see Table 13), while 10 percent live in singletons, and 25 percent in households which they themselves head.

Examining the column "percentage complex," one can see that a sizable minority of women begin their married life in an extended (complex) family and end their lives in one, but that it remains a minority. As desirable as living in a large extended family might be, circumstances have always prevented it from being a permanent solution for the majority of urban residents.

Conclusion to Part Two

Life is difficult for the ordinary Mexican family. It was so even in the years of the economic expansion which ran from 1940 to 1982. The only resource the ordinary Mexican urban household has ever had is itself. Its well-being depends on its ability to exploit itself by organizing income pooling, saving, and consumption in such a way as to maximize the size of the pool and minimize the necessity of counting on the opportunities afforded by economic development.

In the beginning of this section, we sketched our metaphor of the struggle between the individual, the houschold, and the state. Even given that our concept of the state is relatively vacuous in the sense that it is here defined as the organization of power that guarantees the process of accumulation and both foments and mediates struggle within the household largely via the dissemination of the hege-monic ideology and by its attempts at legitimizing itself, we can see that it plays an important role in the daily lives of the Mexican of the city.

The state, and that particular form of economy which Mexico has developed which is sometimes called "dependent capitalism," re-quires a well-disciplined, readily controlled, cheap work force that can be attained in a number of ways: socialization, regimentation, repression, punishment, reward, and cooptation. All of these means are expensive. The more the state relies upon repression, coercion, and force, the more its supervisory powers multiply, and the more the *quis custodiet?* question becomes paramount, perhaps espe-cially in Mexico. Who would police the police force?

Much more desirable is a system in which people permit them-selves to be efficiently exploited, that is, a system in which capital will define the ways that one spends one's working life and the rate at which one will be recompensed. And to serve capital's ends, noth-ing works so well as the constant threat of unemployment or a re-duction in wages (through systematic underemployment), along with the maintenance of an economic system under which the state is seen as responsible for only a part, preferably a small part, of the job creation necessary to absorb people entering the work force. Given the prevailing attitudes toward women, the functionality of the atti-tudes in sustaining the ideology of family relations, the process of social reproduction at minimal cost to the state, and the process of accumulation that the state supervises, it is important that the ordinary people of Mexico reproduce themselves in sufficient num-bers so as to maintain high levels of un- and underemployment.

This is hard to do in a relatively efficient capitalist state that is not totalitarian-fascist, which the Mexican state decidedly is not. People have to be induced to have large families, and the most effi-cient way to do that, while at the same time ensuring their mainte-nance and socialization (for work) at minimal cost, is to make having large families both rewarding to the families themselves and cost free (or thereabouts) to the state. This is what the Mexican system achieves, and why Mexico remains a rich country full of poor people.

What we have called the "Mexican solution" to the problem of poverty is not a solution at all, of course. The underlying rationale of the system puts the individual at odds with his family and house-hold, and the household at odds with the state. A perfectly rational household would retain its members by cutting off their escape routes, which are created through education. Therefore, the per-fectly rational family or perfectly selfish older generation would fail to educate its children past the level where economic opportunity begins to make it possible for the children to abandon their parents, strike out on new career paths, and establish an independent household. The struggle between the collectivist ideology and practice of the household and the individualism that selfishly strives to find a road for each member of the family is thus defined. If the individual "wins" and leaves, the household is the poorer; if the household wins, the state is the poorer because it is robbed of the necessary potential of its workers. And badly educated workers with nowhere to go are not best prescribed for a postcritical Mexico that is opening itself up to the world economy through its entry into GATT.

Thus the dialectic with three terms: individual, household, and state and a dynamic process that must assure that neither gains the upper hand permanently. This is the dynamic that constitutes the ideological field that we examine when we analyze Mexican kinship and domestic organization. But let us suspend this more general discussion until we have had a chance further to examine how the household organizes itself and its resources, and this we do in the third part of this book when we look at how individuals make a living, the role of human capital, the rate of compensation, and the role of formal and informal sectors, and then how households organize their economic resources and budgets for survival.

PART THREE
The Household Economy

In the third section of the book, we analyze the household economy, focusing on the broad outlines of employment and household budget management. In chapter 6 we examine how ordinary people make a living in urban Mexico. We look at the earnings differences between the "informal" and the "formal" sector, finding that the differences in earnings are quite small and that informal sector income opportunities have attractions that might heretofore have gone unrecognized.

In chapter 7 we turn our attention from individuals' jobs and individual labor force participants in the household and look at household incomes and budgets. Before, when we compared the demographic characteristics of the households by economic class, we used the somewhat crude measure of total household income as an index of household welfare and followed the national FOVI guidelines in assigning households of differing incomes to different economic classes. In chapter 6 we take per capita discretionary income (defined as real per household income less expenditures for food, housing, utilities, medical care, and taxes) as our measure to describe the strategies of householders and the way they take advantage of the few opportunities that do exist in the economy. This measure is a much better indicator of the degree of flexibility in household management, both for everyday survival and for potential investments in both children and domestic enterprises. Budgetary flexibility and the ability to respond to crises, challenges, and opportunities is an important goal of domestic budget management among ordinary families, and here, finally, we see some of the advantages of the smaller household. The small household may not be better off in the aggregate, and its overall welfare level may not be as high, but it is better positioned to take advantage of the changing economic and employment conditions of the 1980s.

In examining the household budget and its expenditure categories to see what tactics favor greater flexibility, we discover that the household budget of the ordinary urban Mexican family is geared to expenditure avoidance, rather than income maximization. This does not surprise us, since the opportunities for increasing income are relatively limited, given that the differences between formal sector and informal sector employment are not very large. Even if jobs were available, which they are not, it simply does the householder little good to find another job or to change work sectors. So households must manage with what they have, knowing only that time is marginally on their side, since overall welfare will increase with age, changes in the domestic cycle, and concomitant increases in the earning power of the household. In chapter 6 we develop age-earnings profiles to show that there are increases in income with time (and age and experience on the job), thereby explaining the mechanisms that raised household incomes in our study of the domestic cycle that were examined in the last chapter. In chapter 8 we see that the trick to increasing both flexibility and overall welfare is to avoid spending money on medical bills, entertainment, taxes, and, most particularly, housing. In particular the households that can avoid paying rent and making other housing expenditures are much better off, as renters in both Ciudad Netzahualcoyotl and Oaxaca assured us.

6. Workers, Jobs, and Salaries in Urban Mexico

A householder is defined as a labor force participant if he or she is recorded as earning income, which implies in our survey that they would be assigned to one of the 17 occupational categories ranging from "professional-entrepreneurial" to "unemployed" (see the Appendix for a breakdown of occupational types).[1] In the INDECO survey we distinguished among workers in two ways. The first distinction is between those who hold "registered jobs" and those who have "unregistered" jobs. In Spanish this roughly corresponds to the *asalariado/no-asalariado* distinction. Registered jobs, which are regularly referred to in the literature as "formal sector employment," are more likely to have fringe benefits and some kind of union affiliation. In general, employer-employee relations are more formal, if not legally binding. Second, among unregistered labor force participants, we distinguish those who enjoy the sponsorship of a *patrón*, a relationship that is important to unregistered labor force participants but is not found among formal sector workers. Patron-client relations permeate the whole of the social, political, and economic structure of Mexico in the form of *compadrazgo* relations, relations of collaboration in politics or patron-client relations in the economic sphere (Grindle, 1977). Patrons provide some degree of protection to their client-employees, even though this protection is not bound by the labor codes. Patron-client relations may lack legal recognition, but their binding power may be even stronger. Patrons and clients are often kinfolk, or fictive kinfolk, and patrons are often expected and obligated to act like kin, loaning money during times of need or times of fiesta, attending saints day celebrations, and the like (Eckstein, 1977). It is not unheard of for patrons to register their favored clients in the national social security system.

Those workers who do not hold registered jobs or identify themselves as working for a *patrón* are the unregistered workers. Here we try to identify the informal sector worker, who lacks any kind of contractual guarantee in employment. Few have fringe benefits, aside from those who are enrolled in the social security medical benefits program, which is becoming more and more common in the 1980s, but was not nearly so common when our data were collected in 1977–1979. The unregistered category includes both self-employed and employees, as well as new hires and temporary workers (*eventuales*).

In addition to categorizing workers as registered or unregistered, we also classify them by conventional job type distinctions in four different categories: (1) professionals, heads of enterprises, or entrepreneurs; (2) white-collar workers (*empleados* in public and private agencies and firms); (3) blue-collar workers, or *obreros*, engaged in construction, manufacturing, extractive, and agriculture-related industries; and (4) "tertiary" workers who engage in commerce, services, and artisanal activities.[2]

This two-dimensional classification that crosses the three values of registered status (registered worker, unregistered worker, and worker in patron-client relation) and job type (professionals, white-collar, blue-collar, and tertiary worker) will be employed in the analysis that follows. What we wish to show in the section that follows is that employment can be broken down into segments that offer significantly different earnings prospects. One of the important findings that will emerge from our analysis is that the earnings prospects of registered workers, particularly in the blue-collar category, are not significantly higher than those of unregistered workers.

The formal procedures to be followed involve three steps. First, an earnings model is specified for the job types just outlined, for all the cities of the study. Second, the relative contributions of selected human capital and structural variables (education, occupational role, age, job stability, and city) in explaining variation in individual earnings is examined, especially as they vary across different job and worker categories. Finally, the predicted or expected earnings of actors, based upon individual and structural characteristics, is compared across job and worker categories. Male and female earnings equations are estimated separately, since their prospects vary within the employment market.[3]

Individual Earnings and Worker Characteristics

Our sample of 10 Mexican cities provides a detailed snapshot of workers at all socioeconomic levels during the late 1970s. Tables A.1 and A.2 in Appendix 2 summarize a few of the more relevant details for male and female workers, broken down by broad employment segment and by occupational role. The professional-entrepreneurial category is quite small, representing fewer than 3.5 percent of males and under 2.5 percent of females and therefore not an immediate and practical employment alternative for most Mexicans and consequently is not of great interest in this study. The general category of other white-collar workers, by contrast, represents the largest group, especially among females. This result may be somewhat misleading, however, due to the tendency to classify low-level service people, especially people engaged in precarious commercial activities with the more traditional clerical and administrative personnel in this category.

Males are spread somewhat more evenly among the various broad employment segments, while females tend to be more heavily concentrated in the white-collar category and more sparsely represented among blue-collar workers. Women in the work force have higher average educational levels than men and are, on average, about five years younger than males. Their higher average educational levels may reflect the relatively higher education of younger cohorts within the Mexican population.[4] The female labor force in urban Mexico is different from the male, as Kim (1987) has shown in her study of female labor force participation in these same 10 cities. The female labor force is made up of two segments: a younger segment, mainly unmarried, which has mainly "true" white-collar jobs, and a married, generally older segment that has mainly low-level service jobs. The first group are those young women who have entered the work force out of school with relatively high educational credentials and who will stay there until they get married, get pregnant, or begin to feel they have too many household responsibilities. The second group are those women who are compelled for economic reasons to enter the work force despite their many responsibilities at home. This bimodality in the female labor force makes it more difficult to make general statements based on averages, as we are here, and accounts for the younger average age of women in the work force and their higher educational levels.[5]

Higher educational attainment by women in the work force has not, however, been translated into higher salaries. For all categories, real female earnings were only about 78 percent of those for males.[6] Furthermore, nearly 40 percent of female workers in our sample earned less than the legal minimum wage in their respective regions, as compared to only 30 percent of males. A possible source of distortion in this regard is the unfortunate fact that the data contain no information on hours worked. Thus part-time workers are not distinguished from full-time. With regard to longevity on their current jobs, by contrast, a smaller portion of females seem to have been on their jobs for less than one year—about 20 percent as compared to nearly 30 percent for males. This may be attributable in part to the concentration of females in white-collar jobs, which are presumably somewhat more stable.

Tables A.1 and A.2 in Appendix 2 show that, for both males and females, the distribution of occupational roles within the employment segments differs significantly. White-collar workers are predominantly registered. Blue-collar workers are less likely to be registered, although the majority fall into this category. For tertiary workers the unregistered group becomes the majority. This result is consistent with the general notion that the informal sector is concentrated in commerce and services. These tertiary sector unregistered jobs are likely to be filled by people with less education, and they are less stable as well,[7] as one might expect from the extensive literature on the informal sector.

Unregistered workers are more likely to have no more than primary education, and their average monthly income is lowest. Unregistered workers are more likely to have spent less than one year on the job—as have more than 80 percent of the (unregistered) blue-collar workers and 60 percent of those involved in tertiary activities. In the blue-collar case, this finding indicates that these workers are primarily temporaries (*eventuales*) in both large and small firms. Employment instability for tertiary workers may arise from the fact that such jobs are very easy to get into and out of (by definition). Small business ventures may be lucrative as short-term stop-gap measures during periods of slackness in other job markets, but the risks for the inexperienced are also substantial. This group would seem to be composed of two segments: a more stable group of small-scale operators and a less stable group that consists of those seeking transitory income and others whose small business ventures turned out badly.

The category of "clients," while constituting a relatively small group, provides an interesting contrast with unregistered workers, particularly where average income is concerned. Both male and female clients' earnings are greater than those in the other occupational roles within each broader segment (obviously including registered workers). The average age of clients also tends to be greater, especially in tertiary activities. These patron-client relationships apparently reflect relatively long-standing bonds that are rewarding financially and may also involve varying degrees of protection for clients. The maze of formal and informal patron-client relationships that form the structure of Mexican public and private institutions has been well documented (Grindle, 1977; Lomnitz, 1987; Riding, 1985; Smith, 1979). Such informal relationships contrast sharply with many conventional versions of the "informal sector," especially insofar as it is conceived as the "low-wage" or "unprotected" sector.

A Comparative Analysis of Individual Earnings Prospects

Average figures such as those presented above tend to obscure the existence of substantially different distributions of earnings and the underlying differences in earnings prospects across employment segments and occupational roles for individuals with different sets of resource endowments. This approach requires the introduction of a much more structured analytical framework and the adoption—at least temporarily—of some simplifying assumptions. The approach that follows borrows heavily from Mazumdar's work at the World Bank on Malaysia (Mazumdar, 1981) and from the literature on segmented labor markets, such as, for example, Berger and Piore (1980), Harrison and Sum (1979), and Kerr (1950, 1954).

The logic of the approach to individual earnings determination incorporates certain tenets of human capital theory; namely, that there exists some functional relationship between earnings and both education and experience. The problem is to find adequate measures for education and experience and to estimate the basic parameters of the relationship. If, for example, returns to education are merely the result of credentialism rather than skills related to productivity, then parameter estimates relating education to earnings will be distorted. By the same token, experience may be considered a substitute for formal education. In the absence of data on specific types of individual job experience, age is often used as a proxy for employment experience in general.

Our initial assumption is that education and age are related to earnings, but that earnings are also influenced by such structural factors as employment segment and occupational role relationship. Thus education and age may be considered the basic "human capital" variables in the analysis, while occupational role and a vector of city variables are the basic "structural" elements, subsuming a complicated array of socioeconomic and regional influences. The variable for age is supplemented with an age-squared term under the assumption that the latter will, if statistically significant, account for the presence of so-called age-earnings profiles (Mazumdar, 1981: chap. 6). Viewed from the perspective of the working life cycle from young novice entering the work force to middle-aged experienced worker, these profiles may be interpreted as reflecting the existence of career ladders, learning curves, or seniority rules. Age and experience are important in taking advantage of the limited opportunities for increasing income. With age, one develops a network of contacts (*conocidos*) that can advise one of job opportunities. Second, some job categories commanding higher pay are attained only by experience and reputation: master builder (*maestro albañil* or *maestro de obras de construcción*), for example. Although most Mexicans in ordinary economic circumstances do not think of themselves as having "careers," they do recognize that with age earnings will increase somewhat. At the same time, earnings tend to decline as individuals retire and work less intensively, although the notion of "retirement" is pretty slippery as well. People go on working as long as they can or they have to. Right now there are few *jubilados* or *pensionados* (retirees) among ordinary Mexican families. But still incomes decline, and the curvilinear relation between age and income can be caught with the age-squared variable.

The basic functional relationships of the model may be summarized symbolically as follows:

EARN = f(ED, AGE, INSTAB, OCROLE, CITY), where

EARN is real monthly individual earnings

ED is a vector of dummies[8] for educational levels

AGE represents variables for age and age-squared

INSTAB is a dummy for those with less than one year in their current jobs

OCROLE is a vector of dummies representing occupational role relationships

CITY is a vector of dummies for the different urban areas in the survey

The initial multiple regression estimates are specified separately for males and females and also for the employment segments of white-collar, blue-collar, and tertiary. This breakdown seems justified not only on general socioeconomic grounds but also may be defended statistically, based upon significant results of the so-called Chow-Fisher tests (Pindyck and Rubinfeld, 1981:123–126). In the following tables and discussion, we determine what variables are significant determinants of individual earnings; what portion of the variation in individual earnings may be attributed to these variables;

and how the relative importance of the several variables differs across employment segments.

The least squares regression coefficients for males and females are reported in Tables A.3 and A.4 in Appendix 2 respectively. For males the education dummies are generally significant in all three segments, with the exception of technical training in blue-collar occupations. A likely problem here is simply the small number of workers with formal technical education. As Table 27 indicates, fewer than 2 percent of the blue-collar workers in the survey had technical school training. The basic problem of sample and sample size is especially apparent among females. In the blue-collar earnings equation ($N = 201$), the majority of the dummy coefficients for education are insignificant. For female white-collar and tertiary workers, by contrast, the education variables perform reasonably well.

The coefficients on the age variables are significant for both males and females in each of the employment segments. Age can be viewed as a proxy for experience. The age-squared term estimates the effects of age-earnings profiles. Both are significant for males and females in the employment segments. The age coefficient is relatively low while age-squared is relatively high for male blue-collar workers, suggesting a relatively flat age-earnings profile, at least compared to other segments. For females the opposite is the case, with earnings rising rapidly and peaking at a relatively early age, consistent with Kim's (1987) finding that women enter the work force on roughly equal salary terms with men, but when poverty forces them to re-enter after marriage and children, they receive much lower wages unless they have professional jobs (in which case they are much less likely to leave the labor force in the first place). Table 27 displays the data on educational attainment by sex.

The dummy variable for less than one year's seniority on the current job proved significant in each case except blue-collar females. Blue-collar employment is less of an opportunity for females than it is an option to be avoided. This is particularly true for unregistered blue-collar employment, as the negative coefficient on the dummy makes clear. In the male case the largest negative coefficient is in the tertiary segment. However, the t-value on this coefficient is not significant at the .05 level, a result that reflects the considerable variance within this segment.

Table 28 illuminates the independent explanatory power of each independent variable (or set of categorical variables) with regard to variation in earnings, as measured by the decline in total R^2 as each independent variable is dropped from one of a series of regression equations. As might have been anticipated, education seems to ex-

Table 27. *Distribution of Workers by Highest Educational Achievement*

Highest Education Achievement	Males %	Females %
Primary incomplete	29.9	20.0
Primary complete	30.6	22.1
Secondary incomplete	5.9	4.1
Secondary complete	11.6	17.8
Technical incomplete	0.8	2.3
Technical complete	3.0	10.4
Preparatory incomplete	2.2	2.4
Preparatory complete	3.8	6.3
Professional incomplete	2.7	1.7
Professional complete	9.6	12.9

Table 28. *Relative Explanatory Power of Individual and Structural Variables by Employment Segment and Sex*

	Independent Contribution to R^2		
Variables Excluded	White-Collar	Blue-Collar	Tertiary
Males			
None (R^2, all variables included)	.326	.329	.252
Education	.173	.057	.065
Age and age^2	.071	.044	.067
Under 1 year on job	.003	.006	.008
Occupational role	.001	.010	.018
Urban area (city)	.071	.165	.061
Females			
None (R^2, all variables included)	.235	.294	.341
Education	.140	.047	.075
Age and age ^{2}	.030	.059	.032
Under 1 year on job	.010	-0-	.017
Occupational role	.002	.019	.100

Note: R^2 values correspond to the equations specified in Tables A.3 and A.4 in Appendix 2.

plain the lion's share of earnings variation in the white-collar segment for both males and females. In the other segments, however, the importance of education is less striking. For males in the white-collar and tertiary segments, region and age appear to be roughly comparable. However, region stands out for blue-collar males, as it does for both blue-collar and tertiary females. This result reflects not only the regional diversity of the country but the major impact of key industries. Comparing the regression coefficient for Reynosa with the dummy coefficients on the other city variables in the blue-collar equation in Table A.3 in Appendix 2, for example, illustrates the impact of the petroleum industry on earnings in that city. While the regression coefficients for time on the job and occupational role are statistically significant, as illustrated in Tables A.3 and A.4 in Appendix 2, they do not appear to account for an appreciable portion of the variation in earnings at this level of aggregation.

Regression Results of Human Capital Variables

The foregoing analysis suggests that the determinants of earnings tend to behave differently and that earnings themselves tend to be somewhat lower among the unregistered as compared to registered workers or clients. It should be noted, however, that the approach up to this point has only taken into account horizontal shifts in the earnings function for each occupational role. The coefficients for the human capital variables are only averages for these variables across employment segments. If these coefficients vary significantly among occupational roles, it is necessary to run separate regressions for each role segment to identify these differences. Conceptually, this approach seems justified by the differences in occupational role content. On statistical grounds these separate regressions are justified by the results of the appropriate *F*-tests. Comparing equations containing all three occupational roles with separate equations for each role, we note that the differences in the portion of the variation in real income explained by the independent variables is statistically significant at the .05 level in all cases (Pindyck and Rubinfeld, 1981).

Tables A.5 and A.6 in Appendix 2 present the coefficients for the human capital variables from the earnings coefficients run separately for males and females. In each of the employment segments, the male earnings functions for registered workers explain a greater portion of the variance in earnings than do those for the unregistered group. This result is consistent with the expectation that earnings would be somewhat more structured and consistent in the more formalized part of the economy. The explained variation is especially low in the tertiary unregistered segment, which likely consists largely of independent operators and which conforms as closely as any other to the conventional notion of the informal sector. (Note the R^2 of .174 in Table A.5, as compared to the markedly higher figure for the other categories.)

Unfortunately, the female earnings equations are somewhat less well behaved than are those for males. The registered segments of white- and blue-collar workers are straightforward. However, the educational variables for registered blue-collar workers and for all of the unregistered workers yield rather disappointing results. In the latter case, the poor results may be attributed largely to sample size and distribution. With a sample size of only 47 unregistered blue-collar females, we noticed that observations were missing from several of the educational categories. Consequently, this equation may be judged of little analytical usefulness.

Contributions of Human Capital and Structural Variables

Tables A.5 and A.6 in Appendix 2 offer a breakdown of the variables in the earnings equations in terms of their independent explanatory power, first for men and then for women. Table 29 summarizes the regression coefficients for the structural variables. For both males and females, education makes the greatest relative contribution to explaining the variation in earnings in both the white-collar segment and registered tertiary segment. Age seems to be of greater importance among males than among females, an outcome that may be related to the lower average age of female workers in the sample. And as we suggested above, Kim (1987) has shown, in her study of female labor force participation in these same 10 cities, that the female labor force is made up of two segments: a younger segment, mainly unmarried, which has white-collar jobs, and a married, generally older segment which has low-level service jobs. The first group is composed of those young women who have entered the work force out of school and will stay there until they have too many household responsibilities, and the second group is made up of those women who are compelled for economic reasons to enter or reenter the work force despite their many responsibilities at home. This bimodality in the female labor force makes it more difficult to make general statements based on averages, as we are here.

The large impact of the city dummies on blue-collar earnings, especially in registered jobs, again reflects the impact of key industries. In Reynosa, for example, about 70 percent of the registered blue-collar workers were employed in the extractive sector, which

Table 29. *Relative Explanatory Power of Individual and Structural Variables with Respect to Real Monthly Earnings by Employment Segment, Occupational Role, and Sex*

	Independent Contribution to R^2					
	White-Collar		Blue-Collar		Tertiary	
Variable Excluded from Equation	Regis-tered	Unregis-tered	Regis-tered	Unregis-tered	Regis-tered	Unregis-tered
Males						
None (R^2, all variables)	.338	.274	.367	.293	.299	.174
Education	.185	.108	.038	.077	.110	.040
Age and age^2	.068	.081	.029	.047	.109	.053
Under 1 year on job	.002	-0-	.008	.006	.001	.013
Urban area (city)	.077	.077	.247	.140	.085	.063
Females						
None (R^2, all variables)	.241	.297	.214	.056	.386	.230
Education	.142	.136	.076	-0-	.241	.017
Age and age^2	.030	.007	.081	.056	.050	.019
Under 1 year on job	.016	-0-	-0-	-0-	-0-	.037
Urban area (city)	.062	.074	.038	-0-	.049	.163

Note: R^2 values correspond to the equations specified in Tables A.5 and A.6 in Appendix 2.

consists largely of petroleum workers. The *petroleros* reported average earnings that were roughly double those of other blue-collar workers in Reynosa. This petroleum-based disparity between Reynosa and the other urban areas in the sample is reflected in the relative sizes of the registered blue-collar equation in Table A.5 in Appendix 2. At the other end of this spectrum, workers in Oaxaca consistently stand out as the most poorly paid in the sample.

The variation in both the size of the coefficients on age and education and the explanatory power of these variables across occupational segments may be seen to affect the practical options open to the majority of Mexican workers in numerous ways. For example, the value of education in the sense of an average monthly earnings differential that advances with higher educational achievement, from Mex$2,772 for women who have unregistered jobs in the tertiary sector and no education to Mex$11,378 for men with white-collar registered jobs and a professional level education.

However, for those with neither higher educational credentials nor practical access to them, the relevant question is one of earning a living. From the perspective of those with relatively modest education, access to alternative occupational roles varies significantly from the situation confronted by the better educated. For the 60.5 percent of males and 42.1 percent of females with no more than primary education (see Table 27), tertiary unregistered activities offer some of the more remunerative opportunities.

This is further underscored by comparing estimated age-earnings profiles of workers having no more than primary education across the main employment and occupational segments occupied by the working population. Figures A.1 through A.5 in Appendix 2 present such profiles, derived from the earnings equations presented in Tables A.5 and A.6 in Appendix 2. It should be kept in mind that statistically significant age-earnings profiles derived from regression techniques applied to cross-sectional data do not necessarily establish the shape, or even the existence of, career ladders open to any given age cohort over time. However, the fact that similar profiles may be detected in each of the main occupational roles for males, and in the unregistered tertiary group for females, seriously undermines the conventional notion of the informal sector as little more than a dead-end trap. These profiles represent the modal outlook for the sample in terms of educational achievement.

Examining Figure A.1 in Appendix 2, one sees that for white-collar workers in the primary working years, male earnings generally outpace those of females. For white-collar males, the age-earnings profiles for registered and unregistered workers do not contrast sharply. In the case of females, the apparent differences between registered and unregistered workers must be discounted because of poor quality coefficients. (Note, e.g., the t-value of only 0.88 on the dummy for primary education completed in the white-collar unregistered equation in Table A.6 in Appendix 2.) Due to the same problem of lack of statistical significance, the female blue-collar profiles are omitted from Figure A.2 in Appendix 2. For blue-collar males, the age-earnings profiles track each other remarkably well, with registered workers faring better than their unregistered counterparts by a fairly consistent, though narrow, margin.

In the tertiary segment, the contrast between registered and unregistered workers is not striking in the case of males. For females, the age-earnings profiles track each other closely, but unregistered workers outpace their registered counterparts by a substantial margin (as can be seen in Figure A.3 in Appendix 2). This result is certainly at odds with the conventional formal/informal sector notion that registered employment is generally preferable to unregistered. One possible explanation for this outcome is that it reflects a general employment sphere in which females are systematically discriminated against in registered jobs. In unregistered activities, by contrast, they perhaps enjoy greater latitude.

Unfortunately, as noted earlier, one major obstacle to the empirical testing of a number of propositions relating to earnings differentials between sexes and also among employment and occupational segments is the lack of data on hours worked. If the incidence of part-time work and the distribution of working hours varies across segments, then average monthly earnings comparisons will be distorted. The likelihood of distortions on this account is perhaps greater in the case of females than for males, owing in large part to the complex spectrum of roles filled by women in the context of the household.

In particular, tertiary unregistered roles entered into by women are more likely to function as supplements to household income, complementary to the principal income sources, more than is the case with men. Unregistered females have markedly different family relationships from men and also from other women in and out of the work force. Just under one-half of unregistered tertiary females are household heads (compared to 10% of the sample as a whole), while fewer than one-fourth are spouses of the head. It is highly likely that in many cases these women are operating small shops as supplements to household income, with the assistance of other household members (parents, children, and "outsiders," e.g.). Women involved in such enterprises often spread their time between "business" and "household" activities in such a way that the two tend to merge into one continuous stream that encompasses most of their waking hours and creates the burden of the *doble jornada,* or double day, for them. Earnings from these tiny shops are not regarded in the same way as wages from a factory or a home workshop. These *tienditas* or *misceláneas* are regarded more as a natural extension of household activities in which everybody who is at home engages and in which there is no difference between the stock of the store and that of the larder, as there might be in a proper neoclassically run shop. So, strict comparison between earnings in these enterprises and other kinds of earnings or wages may not be valid. Figure A.4 in Appendix 2 attempts to summarize the contrasts between the tertiary unregistered and the blue-collar registered role segments, since they are the most important job alternatives for men. It can be seen that conventional views of the differences between formal and informal workers are very questionable. Blue-collar registered workers show a more steeply sloped age-earnings profile, with lower predicted earnings at the lower and higher ends of the spectrum. However, relatively lower earnings among younger workers may be explained by more frequent use of unpaid and low-wage family labor and part-time workers in these activities. At the higher end of the profile, lower predicted earnings may be accounted for by the presence of increased numbers of the semiretired, moonlighters, and those who value the advantages of self-employment. Furthermore, for males in registered blue-collar, as opposed to unregistered tertiary, roles, the estimated differential in direct earnings is roughly 7–8 percent, for those with no more than primary education completed.[9] Interestingly enough, the social security benefits (including medical coverage) that are often associated with formal employment are valued, on the average, at roughly 9 percent (Schlagheck, 1980:88–94; Mesa-Lago, 1985).

If the average value of such benefits were added to the mean wages of unregistered blue-collar workers, the result would be almost exactly equal to the average wages of males in unregistered tertiary roles. This finding may yield some insight into the workings of the blue-collar wage mechanism from the standpoint of the implicit supply price required to retain workers in this type of waged employment, given the alternative attractions of independent employment in the tertiary sector.

A systematic treatment of the topics under consideration would be incomplete without reference to the distributions of earnings across segments. At least two reasons underlie the advisability of comparing these distributions. First, from a statistical perspective, treatments based upon comparisons of means—including regression techniques, which are based upon means—are subject to error if extreme values in one distribution or the other tend to skew the resulting statistics. Second, within the context of the theoretical literature, conventional formulations of the formal/informal sector dichotomy would posit markedly different earnings distributions. If the two segments were combined, the expected distribution from this perspective would be bimodal, with "informal" workers tending to cluster at the lower end.

This sort of "dualistic" distribution clearly does not emerge from earnings comparisons of tertiary unregistered males with either their registered tertiary or white-collar counterparts (see Figure A.3 in Appendix 2). While the distribution of the registered tertiary segment is somewhat more peaked than is the unregistered one, the two are remarkably similar. Likewise, these two distributions are quite similar to the registered white-collar group. When we compare the male blue-collar earnings distributions for registered and unregistered workers, the result does lend some credence to the "segmented labor market" view (see Figure A.4 in Appendix 2). However, the comparison of blue-collar registered and tertiary unregistered males shows earnings distributions which are again remarkably similar.

The female earnings distributions presented in Figure A.5 in Appendix 2 offer clearer parallels to the conventional formulations. The white-collar registered distribution is the only one that might be considered quasi-normal. The tertiary unregistered group also exhibits some central tendency and is somewhat skewed toward the lower end of the earnings scale. The tertiary unregistered female distribution offers a clearer contrast to the other primary alternatives than is the case with males. However, as outlined above, the comparison may be obscured because of part-time employment and other considerations.

Finally, as one would expect with earnings distributions, the differences between segments narrows if one compares median earnings, as opposed to means. Mean earnings of 5,715 pesos per month for registered blue-collar workers is 20.7 percent higher than the mean of 4,735 for unregistered tertiary operators. But the 4,000 peso median of the former group exceeds the 3,700 of the latter by only 8.1 percent. In this case the mean figure is skewed toward the higher end of the earnings distribution by the presence of extreme values attributable in large part to the oil workers in Reynosa.

Concluding Observations

What emerges from the data presented in this chapter is that predicted or "most probable" earnings of individuals in registered tertiary roles are not clearly inferior to those in registered blue-collar activities. For many urban Mexican males in the primary working age groups, and with modest (though typical) formal educational attainment, the informal sector may represent an attractive, viable alternative to the poorly paid and monotonous blue-collar jobs which are often viewed with disdain. This conclusion is consistent with the long-recognized fact that self-employment is held up as a goal by many Mexican workers. Whereas independent employment is often viewed as an ultimate goal and a form of social mobility, waged labor is frequently considered as little more than a temporary expedient until adequate resources can be amassed to facilitate independent enterprise (Balán, Browning, and Jelin, 1973).

Having considered the nature of the employment alternatives open to the broad mass of the urban Mexican population from the standpoint of individual earnings, what remains for us to do is to go beyond considerations of individual earnings to examine household strategies not only for earning money income but also for promoting the security of the household in its long-term struggle to meet basic consumption needs.

7. Household Income and Economic Welfare

In this chapter we return to the study of the economic welfare of the household. We think of the economic welfare of a household as depending upon its earnings generated in the employment sphere and its ability to gain access to items of consumption, which is only partially dependent upon money earnings. Many goods and services are obtained without money in urban Mexico, mainly through exchange, or *guelaguetza*, like arrangements with neighbors and kin, or the exchange of hospitality and help with country kin, but also through clandestine means (as in the case of illegal electric lines), political initiatives, or extralegal means of "morally justified acquisition" as, for example, in the invasion of urban land for housing sites.

In this chapter we consider the alternatives available for increasing consumption possibilities by increasing money earnings, and in the next chapter we examine ways for economizing on expenditures. To account for earnings, we focus on the relationship between household income and participation rates in the employment sphere, since the underlying determinants of individual earnings have been examined in chapter 6. This is the input side of the household budgeting equation. In the next chapter we concentrate on the analysis of expenditures, the output side of the budgeting equation, and on strategies for economizing on these scarce inputs to produce acceptable levels of output.

Some Problems in Measuring and Comparing Levels of Welfare

A household, like an individual, has to be able to "defend itself"[1] or to construct shelters against adverse economic circumstances. These shelters or defenses allow some degree of latitude in current expenditure decisions and in a dynamic sense facilitate greater budget flexibility, which may be important as the household's social and economic circumstances change.

Thus household welfare may be conceived not simply in terms of real incomes or real expenditures but also in terms of real consumption possibilities under changing circumstances. The ability of a household to respond to changes in the environment often hinges on the control that it can exercise over basic subsistence needs such as food and shelter, thereby increasing the amount of income in order to "defend itself."

In chapter 5 we used total household income as a rough measure of welfare at the same time that we noted the effect of economics of scale in consumption. For budgetary analysis, as we noted, income or consumption in simple per capita terms as a measure of the relative welfare of individuals or of households could prove misleading, since by weighting all members of the household equally it exaggerated the economic costs of the consumption of children and older people. In this chapter we wish to construct a measure of the "investment possibilities" or the "economic potential" of a household, and for this measure what is crucial is the amount of income that is available for investment (mainly in the children) and the number of people who have claims on that income. So we construct an alternative measure, "per capita residual income," that is, the income that remains after expenditures for the major subsistence items are taken care of. This discretionary income—defined as real household income less expenditures for food, housing, utilities, medical care, and taxes[2]—is a better indicator of the relative latitude that households have above and beyond the bare necessities of life.

In the form of real pesos per capita, the utilization of discretionary income as an indicator of welfare differs somewhat from the focus on per capita consumption in the work of such analysts as Musgrove and Chenery et al. at the World Bank (Chenery, 1974; Musgrove, 1980). This more conventional approach implicitly assumes that consumption expenditures are homogeneous, that they are a generic "good." Concentrating on discretionary income, by contrast, shifts the emphasis from spending per se to the degree of latitude available to household units for spending beyond the purchase of necessities. The latter concept, in other words, is an attempt to direct attention to the strategic alternatives open to families for fulfilling their needs, as they interpret them.

Alternatively, the notion of discretionary income may be viewed as a proportion of total household income, rather than in terms of absolute monetary units. This approach has the obvious drawback of ignoring both per capita considerations and of failing to differentiate between the absolute amount of discretionary income represented by, say, 20 percent of a 10,000 peso income and 20 percent of a 5,000 one. Nonetheless, if gross incomes are segmented by levels such as quintiles, a consideration of discretionary income as a proportion of total income across income levels should highlight key strategic alternatives faced by households that command different resource endowments. In addition, the inverse of discretionary income as a proportion of the total (i.e., the proportion of income consumed by major subsistence expenditures) provides valuable analytical insights into the trade-offs faced by urban Mexican households.

The Options for Raising Household Earnings

To shed light on the dynamic aspects of household budgeting, we need an analytical framework that focuses on the alternatives open to the household to enhance consumption possibilities. One obvious option is to raise total household earnings, a move that is equivalent to increasing total potential household consumption. At the simplest level, the problem may be formulated as an identity. Total household earnings may be conceived as the product of two factors: average income per worker and the number of workers in the household. Although this crude formulation is little more than a tautology, it yields some interesting results. The relationship may be expressed mathematically as follows:

$Y/N = L/N \times P/L \times E/P$, or simply
$Y/N = P/N \times E/P$, where

Y/N is real household income per head

L/N is the ratio of members in the economically active age group to total household members

P/L is the participation rate, in the economic sphere, among household members in the economically active age group

E/P is the average earnings of economically active persons, and

P/N, in the simplified version of the expression, is the actual participation rate, taking into account all household members who are economically active, regardless of age

This expression is useful in that it permits an assessment of the relative contributions of the variation in the two factors that define per capita family income. Since the specification is linear in logs, it can be shown that:

$$\text{var ln } (Y/N) = \text{var ln } (P/N) + \text{var ln } (E/P) + 2\text{cov ln } (P/N \times E/P).$$

The values of the terms in this expression, estimated for those families in the sample of 10 Mexican cities which reported some income and at least one worker, are, respectively:

1. $.5445 = .3254 + .3649 - 2(.0729)$, for all families;
2. $.4223 = .3233 + .3256 - 2(.1133)$, for quintiles 1 and 2; and
3. $.2698 = .3091 + .1681 - 2(.1047)$, for quintiles 3 and 4.

The values of the terms in the above expressions suggest that, for the sample as a whole, variation both in participation rates and in earnings per worker are important in explaining variation in family income. And the relative importance of these two factors (number of members in the work force, average income per member) varies with income level. Average income term is markedly less for the higher income quintiles than for the lower ones. This result underlines the importance of putting more members in the labor force as a way of raising household income, especially for poorer people. The relatively large, negative covariance term, while clouding conclusions somewhat, suggests the extent to which these two factors represent alternatives in the struggle to attain reasonable levels of family income. From the perspective of a household head whose personal earnings are low, the only alternative for increasing household earnings is to increase the number of members in the work force. This is the converse of the inferences from Table 21, which shows that households with more income had more members in the work force and lower dependency ratios despite their larger size.

The P/N term is the inverse of the dependency ratio, or the average number of persons supported by each worker. Keeping dependency ratios down is important. And this is equally true in the poorer 40 percent of the households as it is in the next two higher quintiles. The difference in per capita income between the poorer households of the first two quintiles and the households that are getting by in the second two quintiles is the number of members in the work force, as can be seen by comparing the coefficients on E/P, which is relatively reduced as a contributor to per capita income for the better-off households.

Labor Force Participation Rates and Their Effects

Since the finding of this study concerning the advantages of larger households for economic welfare is so controversial[3] and so contrary to government policies for reduced family size, we hope to be forgiven if we develop the argument in some complexity. First, we present data on household size broken down for more detailed demographic and budgetary analysis.

The size distributions of households show that the lion's share of households falls within the small (42% have between one and four members) and midsized (40% have between five and seven members) range, while large units (with eight members and over) are only 18 percent of the sample. A "typical" household might be said to have four to six members, one or two of which would likely be employed. Single-worker households are clearly in the majority, representing nearly two-thirds of the sample, while 21 percent have two workers, 7 percent have three, and the rest (4%) have more than that. So although a large household with many members in the work force is desirable, it is not very frequent, representing at most around 30 percent of the households. The typical urban household is small or medium sized, with at most two members in the work force.

This seems paradoxical, since it is quite clear that Mexican families and householders are perfectly aware of the ramifications of the argument that we make in this book. There are two reasons for the relative infrequency of the larger groups: first, they are difficult to form and maintain; and, second, there is a competing rationale for the smaller household based on its capacity to invest more in its fewer members. There are three reasons why they are difficult to form: rising age at marriage (Ojeda, 1986), which is delaying the age at which households can be formed; second, the youth of the population in which the median age is around 16, and 35 percent of the household heads are under the age of 35; and third, the very high ratio of migration during previous decades which has made it very difficult for children to form joint households with their parents, who have remained in the villages or towns of rural Mexico. The large households are hard to sustain because of the social and psychological difficulties discussed in chapter 5.

The opposing rationale for the small families is overwhelmingly found in interviews about preferred family size in Mexico, in line with the recent sharp decreases in the urban fertility rate. Our Oaxaca (1987) open-ended interviews showed only one family member (in 28 families) willing to entertain the notion that the larger families were an unequivocal benefit, with no drawbacks; everyone else saw advantages and disadvantages to large families and endorsed the goal of smaller ones. This same ambivalence about large families showed up in the attitudes of men and women who were interviewed separately in the survey of 604 households in Oaxaca, carried out in 1987 as well. An overwhelming majority of the 1,000 men and women who were interviewed indicated that one should think carefully about the consequences and inconvenience of having a baby (91%), while at the same time a minority (38%) endorsed the use of birth control and the notion that one could not be happy without children.[4]

Discretionary Income

There are material advantages to the smaller household as well, since it displays more flexibility than the larger one, by our measure of budgetary flexibility, discretionary income. In Table 30 a "representation quotient" of greater than one indicates that a household size group is overrepresented in a given income quintile, while a value of less than one signifies underrepresentation. At the lowest income level, families of all sizes are represented roughly proportionately. As one moves from the second to the highest quintiles, however, the small households at first are substantially underrepresented and finally are overrepresented. The midsized and large households, by contrast, follow a generally inverse pattern, being slightly overrepresented in the second and third quintiles and similarly underrepresented at the highest levels. The biggest differences are in the second and fifth income quintiles, which are mirror opposites. The second quintile shows an overrepresentation of larger households and underrepresentation of smaller ones, while the fifth quintile has overrepresentation of smaller households and underrepresentation of larger ones. Large households of the second quintile (i.e., large households with little discretionary income) are young and are characterized by high dependency ratios, while the large households of the fifth quintile have lower dependency ratios and are comparatively old.[5]

The advantages of smaller households are illusory in a sense because higher levels of discretionary income are related not only to higher labor force participation rates but also lower dependency ratios. Small households with one member in the work force can lower dependency ratios only by curtailing the number of children, a risky thing to do given the importance of children in the household and the likelihood that they will have to leave home to find work, thereby abandoning the parents and unmarried siblings and leaving them in poverty. And consistent with our earlier finding of the importance of increasing the number of household members in the work force to increase total household income, average (or per capita) participation rates rise by quintile in each household size group, as well.

We have argued the essential rationality, on the part of the poor in urban Mexico, of forming large families as a long-term strategy for relieving, if not escaping from, dire poverty. The consideration of the role of participation rates for families during different phases of the life cycle reinforces this perspective. These life cycle effects can be seen also in the distribution of discretionary income across age-groups and household sizes as Table 31 shows. While large house-

Table 30. *Distribution of Households by Size Groups across Quintiles of Discretionary Income per Capita*

Discretion-ary Income Quintile	Household Size Group		
	1–4	*5–7*	*8+*
1	1.00	.98	1.03
2	.66	1.16	1.42
3	.87	1.11	1.06
4	1.06	.95	.99
5	1.39	.78	.46

Note: Values are representation quotients, defined as the ratio (percentage of households of a given size group in a given income quintile/percentage of total households in that size group).

Table 31. *Distribution of Real Discretionary Income per Capita by Household Size Groups across Household Age-Groups*

Household Age-Group	Household Size Group			
	All Sizes	*1–4*	*5–7*	*8+*
Under 35	.65	.82	.43	.71
35–49	1.04	1.48	1.19	.48
50+	1.75	1.52	2.17	1.78

Note: Values represent income in a given group as a proportion of mean income for the sample as a whole. Age-groups are defined according to the age of the primary female.

holds appear to fare least well in the middle age-group, in the oldest age-group they are second only to the midsized households and are substantially better situated than the average household of any size in the two younger age-groups.

To summarize, the implication of these figures is that family formation over the life cycle may be viewed as a form of investment of human capital—or reproduction of labor power, depending upon one's preferred frame of reference. Provided the household can be kept intact in the later years, sacrifices in the early and middle years due to low participation rates (or, inversely, high dependency rates) may in many cases pay off in terms of greater earnings and added security as the life cycle progresses. Security tends to be enhanced in households with more than one participant (which also tend to be larger households) simply by the fact that earnings do not depend solely on the efforts of one person or, in many cases, even on one occupational type. Multiple household workers with different types of jobs tend to reduce the risk that the consumption unit will be entirely without money income in times of economic adversity. At the same time, there is reason to believe that smaller households, though riskier, are better adapted to changing social and economic conditions, since they are able better to generate more money for investing in improvements, especially in the education for their children, as well as better able to forgo the income that their children might earn were they to leave school.

What we shall discover is that these two competing principles have been worked out and adopted in the households of 1987. Our analysis in Oaxaca, reported in chapter 9, shows that the household that is best adapted to the changing economic conditions represented by the economic crisis in Mexico is a combination of both principles: a large household made up of small families.

Occupational Roles and Household Strategies

In the final section of this chapter, we study employment and household earning strategies and, in particular, follow up the argument about the surprising desirability of informal sector employment as the basis for a household earning strategy, particularly for poorer, older households with lower educational qualifications. We concentrate our attention on three important types of workers: blue-collar unregistered workers, blue-collar registered workers, and tertiary unregistered workers. Other workers (i.e., professionals, entrepre-

Table 32. *Distribution of Households by Selected Occupational Role Groups*

Occupational Role Group	Households %
Single-worker households	
Blue-collar unregistered	7.3
Blue-collar registered	9.3
Tertiary unregistered	13.2
Residual	42.9
Multiworker households	
Blue-collar unregistered	0.7
Blue-collar registered	1.1
Tertiary unregistered	2.0
Mixed	11.0
Residual	12.4

Note: The sample in this case is limited to those households that reported complete occupational data for working members.

neurs, white-collar workers, those holding registered tertiary positions, and all employees of patrons) are simply lumped into a residual category. Units with more than one worker are classified as mixed if they have more than one kind of worker.

Table 32 gives the relative frequency of households with different numbers and types of workers. Unregistered blue-collar employment is apparently only slightly less common than the registered variety, while blue-collar occupational roles are found roughly as frequently as unregistered tertiary employment, particularly when one considers the 11 percent which are mixed (i.e., have workers engaged in two or more of the occupational types identified).

As Table 33 indicates, the average participation rate (first column) in the economic sphere is highest for households with workers devoted exclusively to unregistered tertiary activities, both in the case of single-worker units and those with multiple participants. As Table 34 shows, single-worker households tend to be somewhat overrepresented both in the older age-group and in the small-sized group, as compared to the distribution of the sample as a whole, as well as to the other occupational role categories. Not surprisingly, multiple-worker households devoted exclusively to unregistered

Table 33. *Average Household Participation Rates in the Economic Sphere, Real Income per Participant, and Real Household Income by Selected Occupational Role Groups*

Occupational Role Group	Participation Rate	Income per Participant	Income per Occupation Group
All households	.327	1.000	1.000
Single-worker households			
Blue-collar unregistered	.265	1.055	.916
Blue-collar registered	.247	.738	.606
Tertiary unregistered	.303	.975	.803
Residual	.270	1.324	1.123
Multiworker households			
Blue-collar unregistered	.398	.272	.483
Blue-collar registered	.393	.541	1.011
Tertiary unregistered	.532	.310	.661
Mixed	.472	.548	1.037
Residual	.472	.652	1.186

Table 34. *Distribution of Households by Age-Groups and Size Groups across Occupational Role Groups*

Occupational Role Group	Age-Group		Size Group	
	Under 35	Over 50	1–4	8+
All households	45.0	20.1	41.6	17.5
Single-worker households				
Blue-collar unregistered	51.2	14.8	39.9	18.1
Blue-collar registered	53.3	13.0	39.8	15.8
Tertiary unregistered	44.2	24.5	46.2	13.9
Residual	53.9	13.1	46.9	12.3
Multiworker households				
Blue-collar unregistered	14.5	29.0	24.2	38.7
Blue-collar registered	19.2	32.3	22.2	40.4
Tertiary unregistered	22.3	36.6	35.4	23.4
Mixed	23.4	33.6	25.7	35.3
Residual	34.1	27.3	33.9	25.7

Note: Age-groups are defined according to the age of the primary female.

tertiary activities are concentrated even more heavily in the older age-group and are somewhat overrepresented in the large-sized group. However, these household units are less heavily concentrated in the large-sized category than are the other multiple-worker configurations. This result perhaps reflects the supplementary role of earnings derived from small-scale commercial and service activities, as well as the wisdom of diversifying earnings sources in order to reduce the risk of totally interrupting household income streams during periods of economic adversity.

Aside from the higher average earnings of both individual participants and of household units within the residual category in Table 34, the most attractive employment configuration from the perspective of real monetary earnings is the multiple mixed group. There is a good deal of variation within this group, revolving around the precise employment configuration of these units. Nonetheless, the performance of the "mixed" group underlines the importance of unregistered activities not only as primary employment alternatives but as sources of supplementary earnings from secondary workers. In many cases small-scale commercial and service activities may be seen as valuable income supplements that often make the difference between covering basic household expenditures and living in dire poverty.

Although average earnings per participant are relatively low in households involved exclusively in unregistered tertiary activities, this result does not necessarily indicate that all or even the majority of those who occupy such roles are "underemployed" or represent an "inefficient" allocation of resources. Among the many underlying influences in this regard, the following should be kept in mind. The analysis of chapter 6 shows that workers with minimal formal educational achievement tend to be overrepresented in such occupational roles. On average, opportunity costs to these workers in terms of forgone earnings in, say, registered blue-collar employment, may be quite low.

One problem with comparing the raw earnings of individual workers is that the survey data make no distinction between full-time and part-time workers. Nevertheless, it may be safely assumed that some of the workers employed in unregistered tertiary activities are not strictly full-time. It seems likely that a certain amount of labor is absorbed into such activities which either would not be employed elsewhere in the economic sphere, or would be engaged in some less desirable activity, in the absence of such household operations. The relatively high average participation rates exhibited by households with members engaged in tertiary unregistered activities (see Table

33), especially the overrepresentation of female spouses in these occupational roles, are consistent with this interpretation.

Of course, this perspective does not deny the implication that in
many cases the "opportunities" offered by such activities, in terms
of earnings, may be seen as an onerous form of exploitation, insofar
as individuals may be required to play multiple roles. On the one
hand, the "double day," or *doble jornada*—when wives work outside the home and inside too as housewife—may thus be seen to
have become a "never-ending day" when they must also work as pro-
prietors of small-scale businesses. But such arrangements often employ children and other household members whose contributions,
however marginal, might otherwise go unrealized.

Household Earnings Strategies: Summary

In general, the data presented above may be seen to suggest a pat-
terned set of strategic alternatives for gaining access to levels of
household earnings deemed adequate to cover expenses. For those
units lacking in formal education or some extraordinary means of
access to secure and relatively well paid employment, fielding multiple workers is an obvious and apparently widespread option. Thus
the easy assumption that "the small family lives better" is far too
simplistic. Obviously, the small family supported by one or more
well-paid workers would be expected, other things being equal, to
live better than, say, the large family dependent upon a single low-
wage earner. However, as larger families mature, many may be able
to compensate for lower average earnings by substituting "quantity"
for "quality" of workers—and by diversifying their "investment"
portfolios with regard to the generation and deployment of human
capital.

A final note may be in order concerning further analysis of the determinants of "participation rates." Although a formal, statistical
analysis of the determinants of participation rates might well yield
some interesting results, it would not seem appropriate in the context of the current study. In particular, given the present conception
of an employment sphere that encompasses both waged work in a
labor "market" and a broad spectrum of other activities, not all of
which produce direct money payments, conventional notions of participation as involving a dichotomous "in" or "out" juxtaposition of
individuals vis-à-vis the labor market is of dubious value. While factors such as local unemployment rates, educational levels, wages,
and the like may influence participation in Mexico, as they seem to

do in the United States and elsewhere, their significance is not the central focus of this study. For the majority of working-class Mexicans, the choice is often not one of participation *versus* non-participation, but rather one of what *level* of participation is available and desirable from the standpoint of meeting basic needs.

8. Household Budgeting Strategies

The analysis up to this point has focused on options for raising earnings, but earnings per worker or per household are not the whole story about relative living standards. The way households economize on earnings is just as important. In fact, money earnings do not provide a very reliable guide to living standards in the absence of knowledge about budget outlay,[1] especially the extent of discretionary control over such outlays. In the analysis that follows, we shall focus on budgetary options as they relate to discretionary income per head. We shall question how, with given levels of money income, households may most effectively translate those funds into acceptable living standards by gaining a measure of control over basic budgetary outlays—or, conversely, by increasing discretionary, as opposed to total, income.

This is not to say that consumption of necessities such as housing or health care are not "goods." The point is rather that for poor households "avoiding expenditures" of necessities is both rational and widely practiced. By doing this the household is afforded a wider margin of security against economic adversity. Avoiding expenditures on necessities also enhances the value of "permanent income" as defined by Milton Friedman. In Friedman's conception, current consumption expenditures are a function of "permanent," as opposed to "transitory," income, the former being conceived as the expected value of an income stream extending into the future. The basis for the expected value of permanent income is one's past experience, human capital, and so on, and excludes windfall income and other unexpectable earnings. By controlling expenditures the Mexican urban household can balance its everyday expenses and its permanent income, reserving whatever excess occurs ("transitory income") for housing improvements, new clothing, entertainment, or most important, unexpected educational expenses for the children.

Since risk management increases the expected value of income, one task of household management is the avoidance of risk.

In order to understand some of the budgetary trade-offs, it is useful to consider two major sources of variation in discretionary household income. As a first step in this direction, a relationship of interest may be expressed as:

$$(Y - Es)/N = Y/N \times (1 - Es/Y), \text{ where}$$

$(Y - Es)/N$ is real household per capita discretionary income, defined as real income less expenditures for subsistence, divided by the number of household members

Y/N is simply real household income per capita, and

$(1 - Es/Y)$ is the proportion of total household income available for discretionary uses, after subsistence expenditures are covered. Es/Y is a modified version of the conventional notion of average propensity to consume, and might be denoted the average propensity to consume basic subsistence goods and services. In the context of the present analysis, this variable might be conceived as a "budgeting factor," yielding a measure of the adequacy of household budgets.

The above identity focuses attention on the two major factors that determine the level of household discretionary income per capita. The anticipated relationships—also verified empirically—between the left-hand term and the right-hand factors is positive in the case of Y/N and negative for $(1 - Es/Y)$, since the proportion of income spent on necessities varies inversely with income. (The proportion of income available for "discretionary" uses varies directly with income.) The expression allows a comparison of the relative contributions of the two factors to variation in discretionary income. Since the specification is linear in logs it follows that:

$$\text{var } \ln(Y - Es)/N = \text{var } \ln(Y/N) + \text{var } \ln(1 - Es/Y) + 2\text{cov } \ln(Y/N) \times (1 - Es/Y)$$

The values of the terms in this expression, derived from the sample of 10 Mexican cities, are, respectively:

1. $2.0053 = .7068 + .6991 + 2(.2997)$, for all income classes;
2. $1.4054 = .3200 + .9046 + 2(.0904)$, for the first and second quintiles of total real household income, and

3. $.8720 = .2584 + .4410 + 2(.0863)$, for the third and fourth quintiles of total real household income.

It is worth noting that for the sample as a whole (i.e., for all income classes) the variance in total per capita income is virtually equal to the variance in discretionary income per head (the "budgeting factor"). However, the picture is somewhat clouded by the relatively large covariance term. In the case of the lowest two income quintiles, by contrast, the picture is clearer. The "budgeting" factor is over twice as large as the income variable, even taking into consideration the covariance term. The point to be emphasized is that the *ability to exercise some degree of control over major subsistence items in the household budget is of crucial importance in determining overall levels of household well-being, apart from money income per se.*

The underlying determinants of the budgetary factor are likely to be highly correlated. For example, while food might be considered to take precedence over the other expenditures, food expenditures may be reduced to take care of other pressing needs. Young, low-income households with high dependency ratios and access to cheap housing may spend more freely on food, formal medical treatment, or more "discretionary" goods and services, whereas households paying high rents may have to cut back in these areas.

Table 35, which presents the distribution of real peso expenditures on the five basic expenditure categories, offers some insights into the underlying dynamics of common expenditure patterns. Food and utilities represent standard out-of-pocket expenses, the former commanding by far the largest portion of income. The proportion of households reporting positive monthly expenditures for housing, health care, and taxes, by contrast, is less than 50 percent in each case. Fifty-four percent reported paying no taxes, while 59 percent claimed no monthly housing expenses, and 67 percent reported no health care outlays. Clearly, the key to whether or not many households are able to enjoy tolerable levels of discretionary spending power, despite relatively low levels of money income, may lie in the avoidance of such basic expenditures.

Table 36 presents food and utility expenditures as percentages of total household income, broken down by income quintiles. Not surprisingly, the food burden is heaviest at lower income levels, averaging over 60 percent in the bottom quintile. However, food expenditures are subject to considerable manipulation, allowing many individual households a good deal of latitude either to express their preferences or to cut back in the face of adversity, as circumstances

Table 35. *Upper Limits of Real per Capita Expenditures on Basics (Real Mexican Pesos per Month)*

Quintile	Food	Housing	Utilities	Health	Taxes
1	$300 (US$13)	$0	$30 (US$1.30)	$0	$0
2	465 (US$20)	0	45 (US$2)	0	0
3	630 (US$28)	20 (US$1)	70 (US$3)	0	5
4	915 (US$40)	150 (US$7)	115 (US$15)	40 (US$2)	25 (US$1)

Table 36. *Mean Expenditures on Food and Utilities as a Proportion of Total Household Income*

Income Quintile	Food %	Utilities %
1	61.8	10.3
2	54.9	7.2
3	49.9	6.1
4	38.8	4.4
5	24.7	2.9

allow. In other words, to a much greater degree than the other major expenditure categories, food expenditures exhibit a relatively consistent relationship to income and household size. Consider, for example, the following functional relationship:

$E^f = a + bY + H^n$, where

E^f is the natural log of real food expenditures

Y is the natural log of total real household income, and

H^n is the natural log of household size.

The coefficients corresponding to the independent variables in the above equation, estimated from the sample with ordinary least squares regression techniques, are:

$$E^f = 4.37 + .344Y + .242H^n$$
$$(t = 51.59) \quad (t = 36.33) \quad (t = 13.69); \quad R^2 = .179,$$

indicating that food expenditures are sensitive both to household income and to household size.

As indicated above, the majority of the sample reported making no regular expenditures for health care, shelter, and taxes. Although health care might conceivably vary with household size, no statistically significant relationship could be identified. To the extent that households do incur regular out-of-pocket expenses for shelter and taxes, the amounts are not likely to vary substantially with marginal changes in household size, at least not for low-income households in the short run. By the same token, while most households reported regular expenses for utilities (which include lighting and some form of fuel for cooking and heating), these items are not likely to vary substantially with marginal changes in household size. As a result, food expenditure might be considered as the primary "variable" cost in the household production budget, at least in the short run. The other major subsistence expenditures may thus be conceived as relatively "fixed" factors.

Figure 3 plots the estimated values of total, average, and marginal expenditures for food, derived from the preceding regression equation. Household income is held constant at the median value of some 5,600 real pesos, or approximately US$246 per month. In line with the arguments advanced in chapter 5, what stands out is the relatively low marginal cost of additional household members. At the median household size of five, for example, the estimated marginal cost of an additional member is roughly 100 real pesos per month, or about 3 percent of the minimum salary in the Federal District at the time of the survey. One might add that the proportion of income spent on food does not increase with household size, either, being 44 percent of the budget for small (1–4 members) households and 47 percent for households of five or more members.

It would seem that, for the bulk of the urban Mexican population, the marginal cost of an additional household member, in terms of basic subsistence, is minimal. From the perspective of individual household heads, this result underlines the rationality of investing in "human capital" in the form of offspring, assuming future benefits in terms of financial security and supplemental income in later years. As long as the costs associated with such reproductive activity are low compared to the potential benefits—both monetary and nonmonetary—it is rational to do so. Alternatively, if the household is conceived as a producer of cheap labor power for the indus-

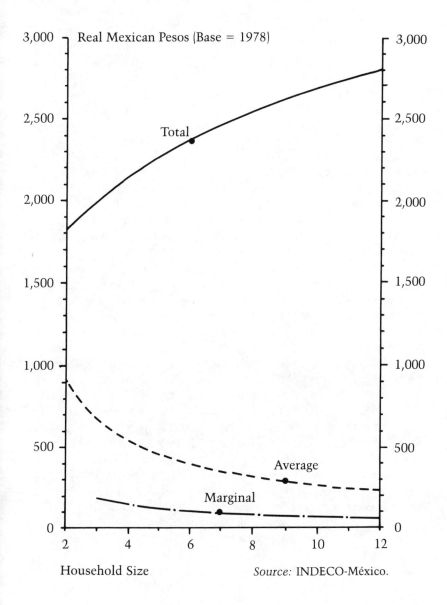

Figure 3 **Estimated Real Monthly Food Expenditures (Median Income Households)**

trial sector, household budget management can be thought of as a technique for doing this as well.

Both perspectives on the working-class household are further supported by the INDECO data on food expenditures across household size groups. The proportion of income expended on food does not increase with household size, which is another indicator of how additional household members become earning assets.

Table 37 highlights the crucial role played by housing and health care expenditures in determining the outcome of the budgetary process. For those household units that report some positive monthly expenditures for either shelter or health care, the budgetary burden is especially onerous at the lowest income level. As would be expected, the burden declines substantially as income rises. However, the proportion of households reporting positive health care expenditures rises steadily with income. This result is consistent with the expectation that adequate health care is something of a luxury that is denied to many poor households.

Access to health care is often associated with the enjoyment of socioeconomic benefits (*prestaciones*),[2] or fringe benefits. The one benefit that is provided if any is is the *Seguro*, or the right to emergency care and minimal hospitalization at the local Social Security Hospital, where the level of care is quite high and is provided through the Mexican Social Security System (*El Instituto Mexicano de Seguro Social*, or IMSS). Almost all salaried or registered employment brings *Seguro*. Close relatives of people with *Seguro* may also gain access to it, and it can be purchased, although the price is usually well beyond the means of ordinary Mexicans and is confined to upscale white-collar workers and professionals. In the survey data, slightly less than one-third of the households reporting no registered workers claimed to have it. By contrast, in those households reporting at least one registered worker, the incidence of benefits was over 70 percent. The lowest incidence occurred among families where no workers could be identified that were engaged in occupational roles other than tertiary unregistered. Not surprisingly, the incidence of such benefits tends to rise markedly with income levels, as indicated by Table 38.

Fringe benefits offer a degree of protection from the financial costs of sickness and injury, and we find that at both lower and higher income levels health care expenditures tended to be a smaller portion of income in households with them than in those without. For households of all income levels that incurred positive health care costs, those without fringe benefits reported expenditures averaging

Table 37. *Households Reporting Positive Monthly Expenditures for Housing and Health Care and Average Income Expended*

Income Quintile	Housing Households %	Housing Income %	Health Care Households %	Health Care Income %
1	33.8	18.9	25.0	11.2
2	37.8	15.5	33.2	7.4
3	42.5	13.8	34.6	6.4
4	43.5	11.9	32.8	4.9
5	45.2	9.4	41.5	3.6

Table 38. *Households Reporting Benefits and Positive Health Care Expenditures*

Income Quintile	Benefits %	Expenditures %
1	36.2	16.5
2	53.8	21.9
3	61.5	24.8
4	68.4	22.8
5	72.4	34.8

7 percent of total income. By contrast, for those units with such benefits, health care expenditures averaged 5.5 percent.

As Table 37 indicates, the proportion of households reporting positive expenditures for health care was related directly, while the percentage of income expended was related inversely to income level. One important consideration in interpreting the behavior of health care expenditures is the fact that, even for those enjoying access to public facilities, these services are not costless. While charges might in general be judged modest by U.S. standards, they are nevertheless significant from the budgetary perspective of many poor households. Consequently, although access to the system does seem to facilitate increased consumption of health care, this advantage involves a regressive expenditure burden.

Housing expenditures are perhaps somewhat more straightforward than those for health care, since the latter is occasioned by unforeseeable and uncontrollable factors such as sudden injury or illness. Housing costs can often be controlled by simple planning, subject, of course, to constraints imposed by the local environment. The quest for home ownership has been an important factor in the growth of working-class housing developments on the periphery of major cities in Mexico and elsewhere. It should be obvious at this point that the success of a household in gaining access to housing at little or no cost, either by legal or extralegal means, can be the decisive factor in determining whether the budget is adequate to cover basic expenditures.

In Mexico people everywhere want to own their own home and lot, as part of the process of "defending themselves." Mexican cities are famous for their undifferentiated low-rise vistas, so much so that one minister of Housing and Urban Settlements was moved to describe Mexico City before the earthquake and despite its high-rise buildings with the colorful descriptions of *"una ciudad, chata, chaparra y cacariza"* (a flat-faced, squat, and pock-marked city—the last because of the many half-completed building sites). A survey of 604 households carried out in Oaxaca in 1987 confirmed that the ideal housing situation for over 90 percent of the Mexicans surveyed was a separate house, which they reported as both the ideal way to live and the way they intended to live.

As Table 39 illustrates, total home ownership is relatively constant across income levels, characterizing about two-thirds of the households surveyed. Irregular tenancy is somewhat more common at lower income levels, with legal ownership tending to increase at higher income levels. The percentage of legal owners bearing positive monthly expenditures for housing is somewhat higher than is the case with irregular tenants (see Table 39). For the latter group the proportion of households reporting no monthly housing expenses tends to decline as income level rises. An important factor underlying this result is that a greater proportion of irregular owners had resided in their current locations for a relatively short length of time (i.e., less than five years in Table 40). At the lowest income levels roughly one-third to one-half of the irregular owners had been in their current locations for under five years. What the data suggest in this regard is that for the poorest households, irregular tenancy affords the opportunity of at least some sort of shelter without the necessity of regular monthly expenditures.

In addition, irregular ownership can often be converted into regular ownership by an administrative procedure, which though somewhat costly and very time-consuming is still economically tolerable and is usually carried out through the auspices of the National Com-

Table 39. *Distribution of Households by Type of Tenancy (In Percentages)*

Income Quintile	Legal Owner	Irregular Owner	Legal Rental	Irregular Rental	Borrowed & Other
1	42.0	26.0	20.1	5.4	6.6
2	43.1	21.0	24.8	4.3	6.8
3	44.0	22.3	25.5	4.9	3.4
4	47.8	19.6	26.9	2.7	3.0
5	51.9	15.7	28.7	2.4	1.3

Table 40. *Households Reporting No Monthly Housing Expenditures (In Percentages)*

Income Quintile	Legal Ownership		Irregular Ownership	
	No House Payment	Under 5 Years	No House Payment	Under 5 Years
1	89.7	17.4	83.8	45.9
2	86.8	18.5	84.9	39.0
3	83.6	18.6	74.3	30.1
4	84.8	18.5	62.6	29.7
5	83.3	16.4	61.4	30.5

mission for the Regularization of Housing Tenure (CORRET). In the majority of cases, it is not necessary to pay a former owner the full cost of land plus improvements, since a settlement can be made based on the value of the land at the time of occupation, which, lacking improvements and "urbanization" of any kind, was usually quite low. Irregular ownership is chancy at first but becomes less so in time. After five years it is highly unlikely that some kind of arrangement cannot be made. Table 40 notes that slightly over 50 percent of the irregular tenancies have gone on that long. Those households that are somewhat better off, by contrast, are able to expend more on shelter, either in terms of increased initial cost or by way of periodic improvements as the flow of earnings allows, as Musgrove (1978) has shown for the ECIEL household budget data for five Latin American cities.

The budgetary burden of housing expenses, for those households reporting such expenditures, is low by Euro-American standards with a maximum of 21 percent for "regular rentals" of the first

income quintile and a low of 6.6 percent for "irregular rentals," averaging approximately 15 percent for all reporting. One factor that
stands out is the very small differences among different types of tenancy. While the housing expenses for householders who rent ap-
peared to be slightly more burdensome for the lowest income households, the differences were minimal. Still, renting is viewed as very
unsatisfactory by urban Mexicans. Half the "migrants" to Ciudad
Netzahualcoyotl did not come from the famished provinces, as so
many denizens of the Federal District believed, but rather from the
neighboring working-class barrio of El Tepito and the surrounding
delegation of Venustiano Carranza precisely to escape the rents
there and to realize their ideal of their own home in a detached
house with a wall around it. They did so despite the fact that Ciudad
Netzahualcoyotl was relatively lawless at the time, completely lacking in any kind of urban amenities like roads, running water, public
transport, and schools and despite the attractions of the old neighborhoods where some of them had lived for generations and which
enjoyed most of the conveniences of urban living[3] (Vélez-Ibáñez,
1983). Of all the bills that the poorer people of urban Mexico dislike,
it is the most detested.[4]

The major problem with our "expenditure avoidance approach" to
the study of household budget and welfare is the lack of any measure
of quality. Of course, conventional "consumption expenditure" approaches have similar drawbacks, insofar as dwellings of comparable
cost are not necessarily of equivalent quality. Table 41 attempts to
shed some light on this problem. Not surprisingly, houses account
for the great majority of dwellings in which tenants have ownership
rights, whether legal or illegal. Those residences classified as shacks
(*jacales*) are somewhat overrepresented among irregular, as compared to legal, owners. Lower-quality residences may be more common among irregular homeowners than among the legal ones. Also,
urban Mexicans, especially poorer ones, build their own homes, and
this takes time—particularly if they have to wait upon regularization procedures and the accumulation of transitory income to pay
for improvements, as well as the time to do them. Building and re-
building a home is, as often as not, a family affair, and scheduling
time for adult family members, all of whom work full time, even
if they are not in the labor force, takes time. *Jacales* are the first step
in the long-term project that eventuates in a solid multiroomed,
sometimes multistoried brick or concrete house.

In the survey householders were asked to rate the quality of their
housing, and irregular owners and renters rated it as lower quality
than regular owners and renters (with 48% of the latter rating it as

Table 41. *Households by Type of Housing and Tenancy (In Percentages)*

	Type of Housing			
Type of Tenancy	Shack	Room	Apartment	House
Legal ownership	5.2	6.1	5.6	83.1
Irregular owner	15.8	12.7	1.6	69.9
Legal rental	3.2	20.3	25.2	51.2
Irregular rental	8.5	33.7	8.5	49.3
Borrowed and other	16.2	25.6	31.7	52.2

either "mediocre," "bad," or "very bad" compared to 60% for the irregular owners or renters). However, irregular owners rate their housing about the same as regular owners and better than irregular renters (70% of whom rated their housing as unfavorable). Furthermore, recalling that roughly three-fourths of irregular owners avoid the necessity of regular monthly housing expenditures and that the average proportion of income expended for shelter by the remaining one-fourth is much lower than is the case for renters of any type, irregular ownership may still represent the best deal available per peso expended. For low-income households, it may represent the preferred alternative.

Property taxes represented an additional expense closely associated with home ownership. Some 65 percent of total home-owning households in the sample reported paying such taxes, the proportion differing markedly between those with regular and those with irregular tenure. Our data indicate that the former were roughly twice as likely to pay property taxes as were the latter. Furthermore, for those regular owners who did report paying taxes, the burden was about twice as heavy as was that of their irregular counterparts, in terms of the proportion of income expended. Although the property tax burden was apparently not great for most owners of either type, irregular owners tended to enjoy an advantage on this account.

Concluding Observations

Clearly, the extent to which households are able to exercise some measure of control over basic subsistence expenditures ranks on a par with money income as a major determinant of living standards. The well-being of low- and middle-income households cannot be adequately assessed by viewing money income in isolation from the consumption side of the budgetary equation, any more than individual earnings can necessarily be inferred from participation in a given employment "sector." While low-cost housing that is substandard is no substitute for raising the inadequate incomes of the poor, the minimization of housing expenditures, both through conventional purchase and extralegal occupation, is an important way of adjusting to the ups and downs of economic life. The latitude afforded by some measure of control over expenditures can make the difference between tolerable hardship and destitution.

9. The Economic Crisis and the New Adaptation

In September 1982 José López Portillo, then president of Mexico, accomplished an astonishing feat. In the last four months of his administration, a supposed lame duck with a successor already elected, with the knowledge of less than six people in his own government, he nationalized the banks. At the same time, he clapped on currency controls, proclaimed himself the scourge of the *sacadólares* (people who had taken their money out of Mexico), and anointed himself Lázaro Cárdenas *redivivus*. The economic crisis that continues until the time of publication of this book was instituted. At first, ordinary people did not care very much. They did not have bank accounts, and so the sight of armed guards around the banks, so emplaced to prevent revolution it was said, did not impress them one way or another. All the ballyhoo about the raising of Lázaro was manufactured as they well knew; some people went so far as to suggest that the president had come off his hinges, lost his spurs as the expression has it, but little suspected that 42 years of stabilized development were coming unraveled before their eyes. When the media's enthusiasm for the immediate political crisis waned, life, in the immediate aftermath, seemed to go on about as it did before. But everything had changed, as was to become clear in the subsequent years. López Portillo was the last of the old-style *caudillos* to inhabit the Mexican presidency. He started his presidency in hauteur and ended it in farce, leaving it for his protégé and successor, Miguel de la Madrid, to pick up the pieces, a favor that the latter would pass along to Carlos Salinas de Gortari.

December 1, 1982, saw the inauguration of the new president. There were warning signs that ordinary people knew that something was wrong. The usual euphoria, laced with strong doses of cynicism, was not to be felt at the inauguration; the crowds were sparse; and the incoming president was too publically bitter about the criminal corruption of his predecessor. Three months after the nationalization of the banks, people were upset: jobs were disappearing; shops were closing; people were restive. For once the fabled Mexican sense of humor was muted; there were no crisis jokes: even the politicians were not funny anymore.

The next years were very difficult indeed for ordinary people in Mexico, as indexed by the fact that their wages were reduced by 40 percent in six years. As the crisis wore on, with inevitable periodic eruptions of euphoria in the newspapers, ordinary people began to believe that the crisis was permanent. As one person we interviewed in 1987 said, "If the politicians think that things have changed for the better, they are lying, because it's not just me that's badly off. . . . none of my friends is doing any better either." In January 1988 the *Economist* would report that Mexico's financial deficit had reached 18.5 percent of the gross domestic product (GDP), surpassing 1982's record 17.2 percent; that inflation had reached 150 percent; and that gross national product (GNP) growth was running at about 1 percent despite the sacrifices of Mexican labor who had suffered a pay cut in every year of the last six. On July 6, 1988, any pretense that Mexico was the same as it had been before the crisis was stripped away when the Partido Revolucionario Institucional (PRI) government had to resort to the crudest manipulation of the election to assure their continuance in power. By June of the next year, even their northern neighbors were sufficiently impressed about the changes in Mexican society and politics to make concessions on the external debt.

Curiously, there were some good aspects to the crisis. Mexicans felt more united than ever before because everyone was affected by the economic crisis. The middle class, which had grown and prospered in the latter years of the great expansion and had become accustomed to taking trips abroad and to the other perks that attend prosperity and an overvalued currency, had gotten their comeuppance; and ordinary people were glad for that. In years past it had galled a lot of ordinary people to see the airs of the new middle class with their television life styles. Ordinary Mexicans felt that they were being cut off from their own compatriots. But now this was changing. Mexicans were all in this together, except for the upper classes, the rich and powerful, and the business people with American friends and bank accounts. They were exempt from suffering despite what the TV program "Los Ricos También Lloran" might suggest. They lived a life apart, but now the term "ordinary" had come to embrace the great majority of Mexicans, perhaps 90 percent of them, and the crisis lent an air of common sacrifice and a bond of unity that the government hastened to exploit. When we interviewed people in Oaxaca about how they were managing in the crisis, sometimes we got the same feeling that one of us had had when ordinary Londoners had talked to him about the blitz, which they had recalled with some pleasure as a time when the class barriers in England had been pierced, as rich and poor, snob and navvy, had huddled in the tube stations together.

What could ordinary Mexicans do to endure the economic crisis? They had no financial reserves to call on. They could not cancel their charge cards, sell the condo on Padre Island, and forgo foreign travel. The obvious strategy would be to do what they had always done: hunker down, call on their families to stick together, and wait out the crisis. Many of them, accustomed to years of slow but steady improvements in Mexico's economic growth (not that they received much of it) were surprised and wondered how it had started, as well as how long it would go on.

View of the Origins of the Crisis from the Top

Since the view is clearer from the top, we can start there to try to understand the roots of the Mexican crisis before we turn to the Oaxaca data to find out what the Mexican household has done to combat its terrible effects. As everyone can now see, the economic crisis did not begin in September 1982. Its beginnings were clearly to be seen in the previous administration's sexennium (1970–1976), and only the discovery of huge deposits of oil and the rise in oil prices had staved off the reckoning that had been better had it come earlier. By the end of the 1960s, the import substitution development strategy had run its course. There were two reasons. First, the import substitution process does have a logic and life course of its own. It starts off by manufacturing the goods that are the simplest and require little or no added infrastructure, nor forward and backward linkages, but as the process moves along, it gets harder and harder because it becomes more complex and involved. At first an entrepreneur could import a piece of machinery, and the inputs, and manufacture a product for sale behind high tariff walls and be guaranteed a profit with minimal competence. But as time went on, more complex and difficult manufacturing processes were required—which themselves require more sophisticated communications, processed inputs, and marketing strategies—and pursuing import substitution got harder and harder. By the end of the 1960s the easy imported products already had their locally manufactured substitutes. The hard, expensive substitutes remained, and these would require higher levels of investment and structural changes in the economy.

The second reason that the import substitution process has a determinate half-life is because of the inefficiencies that it breeds. Locally made products are shielded from outside competition and so can be shoddy and still be sold locally, even though their low quality preempts success on the world market. The policy breeds corruption and large bureaucracies as well. Imports are not only controlled by tariffs but also by licenses and other administrative devices, which provide ample grounds for corruption and bureaucratic control which add to the inefficiency of the process. Import substitution breeds sloppy, fat enterprises selling into captive markets with guaranteed profits. There is a built-in disincentive against exporting goods, except for commodities like oil and minerals and marine products where Mexico has comparative advantage, at the same time that it helps build costly bureaucracies and public sector budgets.

Import substitution policies were part of a larger set of economic relations of dependency that posed additional problems, which were not being attended to in the late 1960s and 1970s. Mexico was not about to, and could not afford to, engage in the kind of research and development that would allow it to produce machine tools and manufacturing processes that were suited to the peculiar conditions of Mexican geography, demography, and economy. It was too difficult, too chancy, and too expensive, particularly when everything could be purchased abroad with little difficulty. This was fine so long as balance of payments difficulties did not intercede, so long as foreign investment and foreign loans made up the differences in the current account. But dependent development has its inefficiencies as well. Payments must be made in royalty form for patents and licenses to foreign firms; profits are repatriated; and the costs of small-scale production for an elite local market must be paid. The call on the surplus is twice what it is in the developed or autocentric economies by some reckoning (de Janvry, 1981), posing great problems for the economy. By 1976 more money was leaving the country than was being brought in: Mexico was being decapitalized, cried the left, as the right made plans to take its capital out of the country.

A third set of reasons for the crisis derived from the Echeverría administration's costly attempt to calm the middle sectors, whose children had revolted in the upheavals of 1968 and 1971. The legitimacy of the system had been radically brought into question, not just the party but the whole state apparatus, the press, and the economic policy of "stabilized development." The disaffected elements had to be accommodated, and they were brought into the party or given some other form of public sector job or handout, once again putting immense and unwelcome pressure on the national budget. (It was said that all but one of the graduating class in the Faculty of Politics at the National University, the most radical group in the uprisings of 1968, had been so accommodated by the early 1970s. Except, of course, for those who had been killed outright, were still in jail, or who had disappeared.) This act of co-optation was very costly and could not be paid out of the oil revenues because there were no oil revenues and would not be for a few years. Rather the oil industry would be borrowing immense sums of money abroad during the period that it was putting in the infrastructure for oil production. From every direction pressures were being put on the public purse, and always there were ample foreign funds to satisfy the demand. Mexico was a good credit risk: 50 years of stable government and 40 years of steady growth averaging 6 percent annually are very appealing to bankers looking for profitable opportunities for Middle Eastern petrodollars.

There were then three sets of causes: the obsolescence of the import substitution model, the turnaround in the direction of foreign funds from net inflow to net outflow, and the costs of legitimizing a regime that had outlived its day. By the middle 1970s a disaster was waiting to happen. In fact, it almost did in Echeverría's last year when the combination of populist adventurism on the part of the president, combined with a hefty and clumsily handled devaluation, induced people to put their money into sounder hands abroad. But the new president calmed the business sector with his probusiness policies called the "Alliance for Production" and the crisis was postponed.

The View from the Bottom

The costs of the inefficiencies engendered by the import substitution process in a dependent economy are paid mostly by labor in the form of reduced or forgone wages. As many people have pointed out, however many arguments there may be about "unequal exchange," and the price elasticities of primary commodities and manufactured goods, and however much it may be disputed whether the developed countries do or do not manipulate the terms of international trade in their favor, there is no argument about relative wages between underdeveloped and developed countries. Wages are lower in Mexico than they are in the United States, about 10 times lower at the time of writing. And low wages are the key element to making the dependent economy work.

How are wages kept down? This book is an essay on the "proximate determinants" of low wages. First, you assure that labor supply exceeds labor demand, and that is primarily a demographic problem. As long as the number of entrants into the labor force exceeds a million, as was true in Mexico in the years 1986–1987, and as long as the number of new jobs created is around 100,000, as it was in 1986, there is going to be downward pressure on wages, except in those areas of high and scarce skills or strong unions where they can be raised by market forces or political forces.

The way that labor supply is kept up is the subject of this book. People were having babies and raising them in Mexico during the 1960s and early 1970s because it made sense to do so. The only way out of poverty for the family, as long as it kept itself together as a pooling and jointly consuming unit, was to insert as many workers into the work force as one could. This was particularly true for the poorest 60 percent of the population (chap. 5) and for the 40 percent who had least flexibility in their budgets (chap. 8). Indeed, we made much of the fact that it was irrational of the householders to invest so much in their children's education and upbringing as to make it necessary for the children to leave home, which would overturn the solution entirely.

But this strategy has changed because of the crisis. Now babies cost more. Food subsidies have been sharply reduced, and the share of food has been increased past the average 40 percent that it was in the late 1970s. (In fact, Lorenzen, 1986: chap. 7, has estimated that for the poorest 30 percent of the people in the poorer cities of Mexico, food budgets by 1986 were taking 80 percent of their household income, a suggestion that was not denied by our interviewing in 1987 in one such city.)[1] Housing costs have risen, if only at the rate of inflation, but the bills are beginning to become a problem of the magnitude that they are to the poor in the United States because the subsidies for electricity and water and other urban services have been cut by providing budgets to city and state governments of the same peso amounts each year during a period of high inflation. The rise in the cost of education has infuriated ordinary people the most. It might appear to be tiny to others, but repeatedly in interviews about the crisis, householders would bring up the added costs of education, explaining that their children could not get any job better than a simple unskilled one (*trabajo sencillo*) without a secondary school certificate, whereas in the past simple literacy was sufficient or, at most, a primary school certificate. At the same time, the direct cost of education has gone up in the form of obligatory contributions to the school budget, the *cooperaciones* and *boletos* that we heard so much about in the interviews. Any schoolteacher can tell you that the budget for the school has declined to the point that not even school materials can be provided: these must be bought by the parents, if they will cooperate. And they do. Education is no longer free, particularly at the secondary level. Credentialism is increasing, and the costs of the credentials are increasing even more.

Employment is down. Without getting into arguments about the degree of articulation of the formal and the informal sector, or indeed about which is the leading sector, or under what conditions one or the other is the source of increase in economic activity, observation and interviews show that informal sector activity is much reduced from 1982 levels, even though more people, especially more women, are employing themselves in these activities. There simply is not as much money around. This means that people do not have their car fixed at the local *taller;* they take their own taco to work with them rather than buy a *torta* at the local *lonchería;* they do not hire people to help finish off the new room in the house; and perhaps most important of all, they do not hire young people as helpers, nor care to purchase goods or services from them as they call from house to house or stand in line among the throng of shoe-shine boys in the *zócalo.*

Children cost more, and they are not able to earn as much as they did in the past. The economic basis for the high fertility rates has been undermined, at least in the cities of Mexico, and the fertility rates are declining rapidly. A closer look at the data from Oaxaca, one of the cities that we studied both in 1977–1978 and 1987–1988, will show in more detail how the people of urban Mexico are adapting to and fighting against the economic crisis that has made their lives so very much more difficult in recent years.

Oaxaca: Ten Years Later

A follow-up study of Oaxaca 10 years later, carried out by Murphy, Earl Morris, and Mary Winter in the NSF-funded project "Household Organization and the Crisis through a Decade of Change," surveyed 604 households of the city in January–March 1987. Fifty households were interviewed in 1987 and 1989 on the topics covered by the survey. We do not claim that Oaxaca is a typical Mexican city; it obviously is not, as its bottom ranking on the "livability" and "urbanization" scales showed in chapter 2. But we do think that a comparison of the condition of the household five years before the onset of the crisis and five years after is not without value and holds some suggestions for how urban Mexicans are faring.

Some Improvements

Perhaps the most important aspect of urban life in Oaxaca is that it has improved in some respects since 1977–1978. Investments in urban infrastructure were sufficient in the precritical period to enable more householders to have access to essential urban services.

Access to the three most important urban amenities—drinking water, drainage, and electricity—has improved markedly in the 10 years. Whereas 48 percent of the households lacked access to running water either in the house, the lot, or in the corner fountain in 1977—1978, 10 years later only 14 percent lacked access. Whereas 68 percent of the households had not had access to proper drainage in 1977–1978, 10 years later that proportion had dropped to 50 percent. Electricity had been unavailable to 28 percent of the households in 1977–1978, but a decade later only 3.5 percent of the households lacked it.

The houses were improved as well. The houses were larger on average than they had been in 1977–1978, although they had about the same number of rooms (averaging 2.5), and the number of houses with more than one floor had almost doubled from 5 percent of the total to 9 percent. Most impressive was the reduction in the number of *jacalitos*, or wattle and daub houses, from 22 percent of the total to 10 percent and an increase in the number of detached or semi-detached houses from 63 percent to 81 percent of the total. Only 7 percent of the households lacked any form of bathroom facility, compared to 23 percent a decade earlier.

Household Composition and Economic Welfare

When the crisis hit, the Mexican urban family could have done two different things, in our opinion. It could have adopted an "everyone for himself" strategy, with every small kinship group seeking its own best advantage without regard to others, and broken up the family; or it could have adhered to the "Mexican solution" to poverty and sought security in numbers and kinship and tried to weather the storm collectively. Because we had been impressed with the fragility of many extended and complex family arrangements, we first thought that the large household would break up. We knew that urban fertility was declining rapidly, and we also knew that the cost-benefit equation that encouraged large numbers of children had changed drastically, raising their costs and reducing the economic benefits that flowed from their work. So we assumed, under an implicit "modernist" paradigm, that Mexico, preparing itself for admission to the international club that GATT built, would be abandoning its traditional extended families of many members and be adopting the small nuclear/matrifocal forms so widely found among poor people in the industrialized countries.

We were quite wrong. The "Mexican solution" continues in force, and the same correlations between economic standing and household organization exist, but even more so. But there is one wrinkle that our previous work should have prepared us for. The most successful household now is a large household made up of multiple small families. Both the welfare advantages stemming from the economies of scale in consumption and income pooling of a large household collective (chap. 5) and the budgetary flexibility of a small family (chaps. 7–8) are to be found in the most successful household collectives of the crisis years. María Jaramillo's family is one like this. About 50 years old, she is an adventurous woman who is fed up with marriage, but keeps her lovers on the place, which is a large lot with three houses on it in a part of Oaxaca that was built up about 15 years ago. She is not always there. When we interviewed the household she was in Culiacán, Sinaloa, a good 800 miles to the north and west, with her three grown but unmarried children working as a *rezagadora*, or sorter, in the tomato harvest. When she is in Oaxaca, she works, as do so many women, going to other people's houses and doing their laundry for very little money, earning a reported 18,000 pesos (roughly US$15) a month. Her lovers contribute to the household's funding as well. But her security is guaranteed by the presence of two daughters (twins, both age 23) and their husbands and the four little children of each family, making for a potential domestic group of 17 people if we count (as they did) María and her three unmarried children as part of the domestic group. The daughters are hard-working laundresses, earning the same amount of money a month as their mother, working with her, and their husbands have more dubious professional pursuits. One is engaged in what one might call light-fingered employment (without victims), while the other is in jail awaiting trial on a rape charge. Their budgetary arrangements are interesting, as one of the daughters explained: "We do not have a common budget for everyone in the house here. Each family eats as it can. From time to time we will eat together when one of us has some important errand to do, but each family is separate. J., for example, (one of the unmarried children of

María) doesn't really get along with his mother, and [he makes it un-pleasant] and that is the reason mother goes to Culiacán." But as she explained later, when an emergency arises, and these are particu-larly frequent in Oaxaca when you have young children, since it is a very unhealthy city, the family is ready to stand by and help finan-cially, morally, and with affection. Another one of the unmarried daughters is a live-in domestic who earns $20,000 a month (roughly US$18) and found, while the son-in-law who is not in jail works at occasional jobs as a mechanic, even though he does not have a proper mechanic's certificate (but "around here the *patrones* don't ask for certificates; they want to know whether you can do the work, that's all"). Extra money is earned by the young marrieds by going around to the houses in the neighborhood on Sundays carrying a huge bag full of used clothing for sale. They think this is a lark and enjoy get-ting away from the kids.

Privacy is a problem on their lot, for although the three houses are separate, what the family has done is built a tiny village in the city. The houses have corrugated steel roofs, dirt floors, and no windows and consist of a single room with a large family bed where everyone sleeps. There are pigs and chickens in the patio.

How do they get along financially? "The money doesn't last. You have to spend a thousand pesos on breakfast, dinner, and supper, and the money's all gone. Our diet is really poor; we have chicken twice a week, and sometimes half a pound of beef jerky (*tasajo*). But for the most part we eat rice, beans, and tortillas, and that's all. We have eggs occasionally, but we never drink milk, because we haven't been able to get a milk card for the children."

They only have themselves for resources. They get up at 6:00 in the morning, have breakfast together (sometimes eggs, often just tortillas and beans), and send one of the young children off to kinder-garten. If there are clothes to wash on the place, then one of the men, if available, goes to fetch water from the corner tap and brings it to the house. The washing begins, if not on site, then off they go to people's houses. She likes to get back by 3:00 in the afternoon so that she can make dinner, but if she is delayed then the man of the house knows how to make fried potatoes, and they keep a day's sup-ply on hand just in case she cannot make it. Dinner is fried potatoes or tortillas and beans, and sometimes eggs. At 5:00 o'clock they go back to work, usually at home, in order to finish the day's laundering by 8:00. At night they sit around talking until the children, one by one fall asleep in their laps and take their place in the bed whence some of them will be ejected in favor of a blanket and a straw mat when their parents retire. There is no television on the property.

Poverty is grueling everywhere, but it is relieved in the case of María's family by the knowledge that they can count on each other. What they fear most is not starvation because food can always be got, and people can live very cheaply indeed. (We found one old man who was living on the equivalent of US$0.35 a day.) What is grueling is the worry about sickness and the children. In the dry season the children get upper respiratory infections, and in the wet season they get diarrheas, and they can die from the conditions induced from either because of bad nutrition and their weakened resistances. And sickness costs money and time, two things that poor people do not have, unless, as here, they are surrounded by family and help.

One should not think that the employment profile in María's family is peculiar, either. The men are essentially unemployed, except for occasional jobs they can pick up around town. The jailed son-in-law was wrongly charged, according to the family; he was no rapist, he was a *vago*, a layabout. The women work outside the house, in unregistered jobs, where in the precritical years they might have been fully engaged working in the home raising the children.

The family as a whole is getting by by living together in separate houses sharing little except fixed housing costs and the occasional meal and family fiesta, thereby combining the two strategies that have emerged as a response to changes in the cost and benefits of children, associated declining birth rates in the city, low employment, especially for males and children, and the need to retain some kind of budgetary flexibility during hard times.

The "Mexican Solution" Continues in Force

A brief look at the relevant data for 1987, comparing it to the 1977–1978 data shows that the "solution" represented by the family of María Jaramillo is typical for Oaxaca and, we suspect, for urban Mexico under the crisis. First, we see that the relationship between economic welfare and family size is maintained. In Table 42 one can examine the mean household sizes for the first three quintiles of household income and the top two, and one finds that the families in the upper quintiles are larger than the families in the lower quintiles and that household size has increased in the period of the crisis from an average 5.26 members to 5.66. Richer households are still larger than poorer ones, and households are still growing.

We found in the precritical period that the better-off households tended to be more numerous and more complex, and the example of María's family might suggest that this tendency would be more pronounced during the period of the crisis, and we are quite correct.

Table 42. *Household Size by Household Income Quintile*

Number in Household	Quintiles 1–3	Quintiles 4–5	Total
1977			
0–4	45%	34%	40%
5–7	43%	43%	43%
8+	12%	23%	16%
N	887	660	1,547
Mean	4.90	5.80	5.26
Median	4.80	5.50	5.00
1987			
0–4	43%	29%	37%
5–7	44%	47%	45%
8+	18%	13%	15%
Mean	5.10	6.20	5.66
Median	5.0	6.0	5.5

Table 43. *Household Complexity by Income Quintile, 1977, 1987 (In Percentages)*

Household Type	Quintiles 1–3	Quintiles 4–5	Total
1977			
Nuclear	80	72	77
Complex	20	28	23
1987			
Nuclear	77	62	71
Complex	23	38	29

Table 44. *Wives in Work Force by Quintile of Household Income, 1977, 1987 (In Percentages)*

Households	Quintiles 1–3	Quintiles 4–5	Total
1977			
Wives not in force	82	61	74
Informal sector	5	26	13
Formal sector	13	13	13
1987			
Wives not in force	66	49	59
Informal sector	30	31	30
Formal sector	3	20	11

Breaking out the households by income quintile, comparing the poorer 60 percent with the better-off 40 percent of the households shows that complexity increases with economic level and over time as well. Table 43 presents the data.

Income-earning patterns have changed as well as a result of the crisis. Although the better-off households still show many more members in the formal sector of the work force (in registered jobs), there have been dramatic shifts in the sectoral distribution by sex, at least from our Oaxaca data. During the period of the economic crisis, women have been brought into the work force in unprecedented numbers, and their work has been overwhelmingly in unregistered jobs, in the informal sector. María's daughters were not at all atypical according to the statistics. Table 44 focuses on working wives and shows both the increased proportion of them in the work force and the sectoral participation rate.

Conclusion: Effects of the Crisis

The family that stays together reduces the impact of the economic crisis. In that sense nothing has changed, affirming the truth of those skeptical neighbors in Ciudad Netzahualcoyotl who said, "Crisis? What crisis? There's always a crisis around here! We have always been in crisis!" The crisis for poor people has been permanent, so their reaction to its conditions does not change, only intensifies.

Individual incomes have been reduced in real terms by over 40 percent since the beginning of the crisis. Household incomes in Oaxaca have been reduced in value *only* 23 percent in the 10-year period covered by our two studies. We cannot say how much of the reduction of the decrement in income is accounted for by increased household size and the new pattern of household organization, but we feel that these changes have had significant income effects.

But the crisis has taken a toll on the family, and some have not survived. It is still "hard living" (to use an American phrase) in the city. Mexican households have always been subject to violence, no less so than elsewhere, and it has frequently been perpetrated by men, drunken men, and drunkenness may have been curtailed somewhat by reduced incomes, but drink can be bought very cheaply. "Only the police can afford their Don Pedro,"[2] they say in Neza, and it is true that ordinary people have switched to *aguardiente,* and drunkenness though less has not entirely disappeared. Whether the violence that makes living on the margin in the city such hell is greater or less as a result of the crisis is a matter of some debate among people who study the effects of the Mexican crisis on Mexican family life. No one has definitive data on this topic, although González de la Rocha's group has harrowing case histories in which women bear testimony to the violent consequences of the crisis in their lives. Their testimony bears witness to the fact that the frustrations of the crisis have been terrible, and in particular, the shift in employment patterns that renders males redundant and introduces women into the work force in great numbers has led to disruptions in the family. Male dignity has been so assaulted by unemployment and the necessity of relying on women for the substance that men formerly provided, that men have taken it out on their wives, and domestic violence has increased. Just as unemployment is a key indicator of increased rates of alcoholism, divorce, and family disorganization in the United States, so too in Mexico some of us feel that family life has deteriorated greatly as a result of the crisis. And the families which have been riven by fighting and brutality can easily be said to be the true victims of the economic crisis. If development could be said to have taken place at the expense of women and children, so more can it be said that the crisis is being managed at their expense.

But there is another side of the picture. Not for many years has each member of the household needed the others so much as they do now. Nothing so breaks up households and leads to divorce and abandonment as domestic violence. Never have men needed their women more than right now; not just to cook, keep house, mend the clothes, and look after the children, but to earn the income that makes it possible to survive. If our argument about the bullying macho getting his comeuppance when his wife leaves him and forms a matrifocal household was correct (remembering Chant's data on the frequency of matrifocal arrangements initiated by the women themselves), then survival can only happen when violence is controlled. It is too optimistic, not to say simplistic, to suggest that the survivors of the crisis are creating a whole new form of Mexican family life, more democratic, more shared, and more tranquil. But it is not too much to say that survival these days requires a wholly new concept of family life, as well as a new conception of the kinds of sacrifices required just to survive. As a 60-year-old man living with his mother-in-law, married daughter, five children, and a grandchild said in an interview we did in 1987:

It was different in the old days. My father brought us up on 130 pesos a week, and we lived all right. Sure, we had to wear hand-me-down sandals, but there was always enough to eat. I didn't even finish primary school and I had to go to find a job outside my family, but I survived. There were seven children in the house, and I was 21 when my father died, and 22 when my mother died. Then I got married.

But you know, men used to have a harder life than women but it's different now. In the past a man didn't want his woman to leave the house to work, but now she has to, because you need two wages just to survive.

Then his wife added: "One has to get used to it. It is difficult keeping the family going, when you have five thousand pesos to buy break-fast [then a little over US$3.00 to feed 12 people]. You have to be so careful not to waste anything. Often we don't have very much to put into the family pot."

The crisis is changing Mexico, and some of the most important changes are occurring from below. The Mexico that comes out of the crisis will not be the society that entered it, and one must hope that the survivors set the pattern for a society in which the concept of "popular" will not be synonymous with poverty and suffering but will serve to define a democratic, liberated Mexico in which the weak and the poor will not be the only firm foundation for the rich and the strong as they have been for so long.

Appendix 1   Study Methods

A more detailed description of the rationale, methods, and approaches of the study can be taken from an earlier project document (Selby and Murphy, 1979) that describes the procedures taken in Oaxaca (see also Murphy, 1979). These procedures were followed in the other nine cities that ultimately made up the study sample.

The study has two phases. The first is a study of the whole city using a method for survey evaluation of urban dwelling environments, developed by the School of Architecture and Planning of the Massachusetts Institute of Technology and described in Baldwin (1974) and Caminos, Turner, and Steffian (1969). The second phase consisted of giving two questionnaires to a sample of the city's population: one concerning the physical conditions of life and amenities enjoyed by the population and the other concerning household membership and the social and economic conditions of the household.

Phase 1

Available information was used to put together a general picture of the city's population, growth, climate, spatial configuration, types of land tenure, and patterns of land use. This information was gathered from government agencies and from "qualified informants" such as local social scientists and politicians who had knowledge of the city. The data were then used to divide the city into "localities" which were "relatively self-contained residential areas within an urban context. In general [a locality] is contained within physical boundaries that are of two types: barriers and meshing boundaries. Mountains, water, limited access highways and sharp changes in land use are considered barriers. Main streets and political or municipal divisions are considered meshing boundaries" (Caminos, Turner, and Steffian, 1969:16).

Because the localities were defined by physical barriers for the most part, air photos and inspection trips were sufficient to identify 24 localities in Oaxaca.[1] Two-person teams consisting of a social scientist and either an architect or an engineer checked each locality to ensure that it was relatively homogeneous. Areas that differed in land or house tenure, housing styles, age of settlement, and the general socioeconomic condition of the population were designated "sublocalities" and noted on a map. There were 116 in all, and they corresponded almost exactly with the political divisions or *colonias* (neighborhoods) of the city. The *colonias* became the units of study for Phase 1. The research team chose within each *colonia* a representative segment of 400 by 400 meters. Detailed notes were taken concerning the most common house types and construction materials, available amenities and services, and the social, economic, and tenure situation in the neighborhood. With this information the survey team chose the most typical block in each *colonia*.[2] While the architect was measuring a home and gathering data on the physical aspects of the house and lot, the social scientist interviewed an adult member of the household and one other in the same block. The interviews concerned the household, its members, their history, and the history and characteristics of the neighborhood.

In addition to providing the data for a general description of Oaxaca, the first phase of the study was designed to develop a typology of distinctive neighborhood types. The classification of neighborhood types reflected the type of house, lot tenure, standard of living of the local people (their income, residential history, and education level), and the desirability of the neighborhood (local organization, morale of inhabitants, social cohesion). The combination of these characteristics led us to define the following eight types of neighborhoods: invasions, *colonias populares* with very low incomes, *colonias populares* with intermediate incomes, *colonias populares* with moderate incomes, site-and-service projects, *pueblos conurbados* (or villages that had been incorporated into the city while still retaining some of their original characteristics), central city neighborhoods, and middle-class neighborhoods.

Appendix 2 Supplementary Tables and Figures for Chapter 6

Table A.1. *Distribution of Male Workers, Average Age, and Average Monthly Earnings by Employment Segment and Occupational Role*

Segment	Workers %	Average Age	Average Earnings in Pesos
Professionals and entrepreneurs:	3.3	39.7	11,845
Registered	1.3	36.3	11,845
Unregistered	1.3	39.6	10,242
Client	0.8	44.9	14,346
White-collar	37.5	36.0	7,028
Registered	32.3	36.2	7,010
Unregistered	4.6	35.1	6,908
Client	0.6	36.2	8,860
Blue-collar	25.2	37.9	5,244
Registered	13.8	37.4	5,715
Unregistered	10.7	38.5	4,550
Client	0.7	40.2	5,796
Tertiary	33.9	38.6	5,446
Registered	14.1	36.9	5,543
Unregistered	17.1	39.2	4,735
Client	2.3	44.3	10,232
Total	100.0	37.5	6,253

Note: Sample size is roughly 9,000, differing somewhat among variables because of missing values.

Table A.2. *Distribution of Female Workers, Average Age, and Average Monthly Earnings by Employment Segment and Occupational Role*

Segment	Workers %	Average Age	Average Earnings in Pesos
Professionals and entrepreneurs	2.3	33.8	7,846
Registered	1.6	33.1	7,371
Unregistered	0.4	35.7	8,785
Client	0.2	35.0	9,125
White-collar	51.9	29.6	5,478
Registered	46.4	29.6	5,538
Unregistered	4.7	29.6	4,746
Client	0.8	31.8	6,674
Blue-collar	9.3	30.9	4,618
Registered	6.9	28.5	5,088
Unregistered	2.2	38.1	2,372
Client	0.1	32.7	9,558
Tertiary	36.6	36.8	3,725
Registered	13.2	31.6	4,244
Unregistered	21.7	39.3	2,897
Client	1.7	44.9	9,902
Total	100.0	32.4	4,890

Note: Sample size is roughly 2,300, differing somewhat among variables because of missing values.

Table A.3. *Male Earnings Equations by Employment Segment*

Variable	White-Collar (N = 3,313)	Blue-Collar (N = 2,204)	Tertiary (N = 2,595)
		Employment Segment	
Constant	2.0756	2.4104	1.9074
	(19.90)	(22.08)	(16.60)
Highest educational achievement			
Primary complete	.1406	.2179	.1990
	(4.80)	(8.02)	(6.45)
Secondary incomplete	.2208	.3071	.2076
	(4.60)	(5.90)	(3.51)
Secondary complete	.3318	.4418	.3695
	(9.61)	(9.98)	(7.84)
Technical incomplete	.6223	.2784	.3851
	(6.31)	(1.72)	(2.75)
Technical complete	.5682	.2236	4.20
	(11.45)	(1.91)	(4.20)
Preparatory incomplete	.4736	.5291	.6504
	(9.86)	(5.10)	(8.04)
Preparatory complete	.466	.684	.280
	(9.60)	(4.56)	(3.91)
Professional incomplete	.5216	.7033	.3930
	(9.77)	(4.75)	(4.15)
Professional complete	.9175	.6536	.9165
	(26.34)	(6.82)	(11.43)
Age	.0775	.0509	.0806
	(14.30)	(9.31)	(13.86)
Age^2	.0008	.0005	.0009
	(−12.08)	(−7.80)	(−12.51)
Under 1 year on job	.1098	.1366	.1594
	(−2.80)	(−4.58)	(−5.07)
Occupational role			
Unregistered	.0785	.1900	.6109
	(−2.23)	(−6.05)	(−1.87)

Table A.3. (*continued*)

Variable	Employment Segment		
	White-Collar (N = 3,313)	Blue-Collar (N = 2,204)	Tertiary (N = 2,595)
Client	.0098 (0.12)	.5238 (−0.64)	.3827 (6.94)
Urban area (city)			
Querétaro	.0259 (0.51)	.0684 (1.23)	.0202 (0.38)
San Luis Potosí	.0228 (0.57)	.0413 (0.83)	.0718 (1.40)
Tampico	.1579 (3.85)	.2816 (4.38)	.2306 (3.87)
Reynosa	.3100 (8.36)	.6033 (12.91)	.3527 (7.89)
Mexicali	.0487 (−1.04)	.1075 (1.82)	.0917 (1.64)
Mazatlán	.2843 (7.19)	.4986 (8.57)	.4278 (7.69)
Oaxaca	.2400 (−6.96)	.1099 (−2.14)	.1763 (−3.80)
Villahermosa	.1632 (2.56)	.3231 (4.53)	.0647 (0.74)
Mérida	.1870 (−3.96)	.0206 (−0.29)	.0470 (−0.77)
R^2	.326	.329	.252

Note: The dependent variable is the natural log of real peso earnings. The constant term, or base, represents Venustiano Carranza (in the Federal District) and workers with primary education incomplete, over one year on the job, and registered jobs.

Numbers are regression coefficients; *t*-values are in parentheses.

Table A.4. *Female Earnings Equations by Employment Segment*

| | Employment Segment | | |
Variable	White-Collar (N = 1,150)	Blue-Collar (N = 201)	Tertiary (N = 679)
Constant	2.7017 (16.87)	1.5986 (3.16)	1.9907 (7.28)
Highest educational achievement			
Primary complete	.2198 (3.38)	.2012 (1.46)	.2792 (3.59)
Secondary incomplete	.2203 (2.37)	.2290 (1.22)	.0239 (0.14)
Secondary complete	.3596 (6.21)	.7957 (4.25)	.4762 (3.69)
Technical incomplete	.4944 (4.92)	.6833 (1.30)	.2200 (0.63)
Technical complete	.5263 (7.75)	.2043 (0.70)	.7496 (3.87)
Preparatory incomplete	.5510 (5.54)	.4719 (1.39)	.8109 (3.19)
Preparatory complete	.5176 (6.88)	.5119 (1.65)	.9662 (4.95)
Professional incomplete	.5403 (4.81)	1.1815 (1.46)	.8482 (2.63)
Professional complete	.7969 (12.21)	.3506 (0.96)	1.1429 (7.28)
Age	.0395 (4.44)	.0972 (3.23)	.0579 (4.11)
Age^2	−.0004 (−3.48)	−.0012 (−2.75)	−.0006 (−3.28)
Under 1 year on job	−.2874 (−4.05)	−.1610 (−1.00)	−.3068 (−4.23)
Occupational role			
Unregistered	−0.445 (−0.78)	−.5050 (−2.58)	−.1546 (−1.79)
Client	.3176	−.1736	.6495

Table A.4. (*continued*)

	Employment Segment		
Variable	*White-Collar* *(N = 1,150)*	*Blue-Collar* *(N = 201)*	*Tertiary* *(N = 679)*
Urban area (city)			
Querétaro	−.1466 (−2.08)	−.1735 (−0.75)	−.2403 (−1.65)
San Luis Potosí	−.0582 (−1.10)	−.1342 (−0.85)	−.1360 (−1.09)
Tampico	.0528 (0.70)	.4754 (1.27)	.2314 (1.65)
Reynosa	.1918 (3.41)	.5946 (2.99)	.4853 (4.13)
Mexicali	.0067 (0.10)	.0638 (0.36)	.1901 (0.96)
Mazatlán	.1792 (3.24)	.8196 (1.93)	.1762 (0.95)
Oaxaca	−.2322 (−4.42)	−.1446 (−0.58)	−.4581 (−4.58)
Villahermosa	−.0084 (−0.10)	−.3385 (−0.84)	.0090 (0.04)
Mérida	−2.861 (−4.47)	−.5198 (−1.92)	−.2610 (−1.92)
R^2	.235	.294	.341

Note: The earnings variable is the natural log of real peso earnings. The constant term, or base, represents Venustiano Carranza and workers with primary education incomplete, over one year on the job, and registered role relationships.

Numbers are regression coefficients; *t*-values are in parentheses.

Table A.5. Male Earnings Equations by Employment Segment by Occupational Role

	White-Collar		Blue-Collar		Tertiary	
	Registered (N = 2,835)	Unregistered (N = 421)	Registered (N = 1,212)	Unregistered (N = 942)	Registered (N = 1,026)	Unregistered (N = 1,396)
Constant	2.1126 (19.10)	2.0012 (5.89)	2.6397 (18.25)	2.3608 (11.43)	1.5471 (9.08)	2.15698 (13.64)
Highest educational achievement						
Primary complete	.1423 (4.55)	.1072 (1.24)	.1789 (4.96)	.2126 (5.13)	.1624 (3.52)	.1826 (4.34)
Secondary incomplete	.2600 (5.11)	.0836 (0.56)	.2065 (3.31)	.4017 (4.34)	.2847 (3.48)	.0651 (0.74)
Secondary complete	.3146 (8.65)	.3926 (3.45)	.3149 (5.91)	.5739 (7.15)	.3713 (5.40)	.3039 (4.49)
Technical incomplete	.6532 (6.22)	.4049 (1.47)	.2492 (1.35)	.1361 (0.36)	.3808 (1.87)	.3595 (1.92)
Technical complete	.5474 (10.85)	.9244 (4.35)	.1437 (1.07)	.0649 (0.27)	.4081 (3.53)	.3441 (2.16)
Preparatory incomplete	.2956 (4.44)	.3806 (1.70)	.3367 (2.86)	.6398 (3.18)	.7869 (5.83)	.2197 (1.48)
Preparatory complete	.4753 (9.38)	.4672 (3.10)	.3540 (3.07)	.9562 (4.05)	.6013 (5.30)	.6842 (5.56)

Table A.5. (*continued*)

	White-Collar		Blue-Collar		Tertiary	
	Registered (N = 2,835)	Unregistered (N = 421)	Registered (N = 1,212)	Unregistered (N = 942)	Registered (N = 1,026)	Unregistered (N = 1,396)
Professional incomplete	.6034 (10.70)	.0012 (0.01)	.5115 (3.11)	1.0329 (3.35)	.5975 (4.89)	.1075 (0.62)
Professional complete	.9050 (25.13)	.9731 (6.98)	.5823 (5.83)	1.2324 (3.16)	.9780 (10.32)	.8299 (5.11)
Age	.0752 (13.06)	.0789 (4.74)	.0410 (5.40)	.0508 (6.35)	.0974 (10.76)	.0694 (8.91)
Age^2	−.0008 (−11.10)	−.0008 (−3.80)	−.0004 (−4.48)	−.0005 (−5.37)	−.0010 (−9.46)	−.0008 (−8.18)
Under 1 year on job	−.1465 (−2.76)	−.0052 (−0.07)	−.1411 (−3.62)	−.1387 (−2.84)	−.1001 (−1.73)	−.1865 (−4.70)
Urban area (city) Querétaro	.0202 (0.39)	.0155 (0.06)	.0877 (1.49)	−.1042 (−.065)	.0391 (0.59)	.0007 (0.01)
San Luis Potosí	.0131 (0.32)	.0190 (0.10)	.0166 (0.32)	.0329 (0.23)	.0350 (0.50)	.1122 (1.32)
Tampico	.1099 (2.50)	.1998 (1.19)	.2201 (2.68)	.2103 (1.45)	.0950 (0.93)	.2247 (3.02)

Reynosa	.3525	.0957	.8326	.3292	.5127	.2402
	(9.17)	(0.58)	(15.81)	(2.50)	(7.16)	(4.12)
Mexicali	-.0676	.0084	.0720	.1451	.0808	.0181
	(-1.41)	(0.04)	(1.17)	(0.84)	(1.16)	(0.19)
Mazatlán	.3008	.3071	.3086	.5414	.5112	.4758
	(7.50)	(1.54)	(4.25)	(3.82)	(4.80)	(5.90)
Oaxaca	-.2196	-.4661	-.2407	-.2007	-.2894	-.1793
	(-6.36)	(-2.55)	(-3.64)	(-1.51)	(-3.65)	(-3.08)
Villahermosa	.1918	-.2795	.3495	.1339	.2035	-.0174
	(3.04)	(-0.68)	(4.05)	(0.85)	(1.66)	(-0.14)
Mérida	-.1353	-.4779	.0229	-.1872	-.1279	-.1105
	(-2.78)	(-2.41)	(0.27)	(-1.30)	(-1.39)	(-1.30)
R^2	.338	.274	.367	.293	.299	.174

Note: The dependent variable is the natural log of real peso earnings. The constant term, or base, represents Venustiano Carranza and workers with primary education incomplete and over one year on their current jobs.

Numbers are regression coefficients; *t*-values are in parentheses.

Table A.6. *Female Earnings Equations by Employment Segment by Occupational Role*

	White-Collar		Blue-Collar		Tertiary	
	Registered (N = 1,032)	Unregistered (N = 105)	Registered (N = 152)	Unregistered (N = 47)	Registered (N = 210)	Unregistered (N = 444)
Constant	2.706 (16.06)	2.7735 (5.34)	1.1353 (1.79)	.7633 (0.40)	1.7769 (4.19)	2.1915 (5.84)
Highest educational achievement						
Primary complete	.2084 (3.01)	.1925 (0.88)	.1483 (0.94)	.2748 (0.72)	.1862 (1.40)	.2802 (2.91)
Secondary incomplete	.2029 (2.10)	1.1014 (2.25)	.2214 (0.91)	N.O. —	.4519 (1.51)	-.2644 (-1.22)
Secondary complete	.4121 (6.11)	.1324 (0.58)	.7556 (3.95)	N.O. —	.6977 (4.02)	.3669 (1.87)
Technical incomplete	.4826 (4.62)	.3715 (0.94)	.5998 (1.17)	N.O. —	-.0796 (-.20)	.6041 (1.07)
Technical complete	.5215 (7.30)	.3187 (1.25)	.0755 (0.21)	.5403 (0.72)	1.1285 (4.79)	.3133 (1.04)
Preparatory incomplete	.4744 (4.39)	.5188 (1.77)	.3617 (0.96)	1.3405 (1.17)	.8544 (3.31)	1.1449 (1.40)
Preparatory complete	.4972 (6.20)	.4517 (1.79)	.4233 (1.39)	N.O. —	1.0618 (5.11)	.5177 (1.09)

Professional incomplete	.6337 (5.27)	-.1936 (-.059)	1.7436 (1.96)	N.O. —	1.2276 (3.03)	.1862 (0.32)
Professional complete	.7859 (11.46)	.8492 (3.27)	.5341 (1.43)	N.O. —	1.2672 (7.25)	.6006 (1.45)
Age	-.0392 (4.15)	.0442 (1.69)	.1340 (3.24)	.1091 (1.46)	.0572 (2.50)	.0489 (2.65)
Age²	-.004 (-3.23)	-.006 (-1.65)	-.0018 (-2.48)	-.0013 (-1.22)	-.0006 (-1.80)	-.0005 (-2.20)
Under 1 year on job	-.4520 (-4.71)	-.0261 (-0.19)	-.0233 (-0.12)	-.1834 (-0.47)	-.92 (0.05)	-.3951 (-4.60)
Urban area (city)						
Querétaro	-.1583 (-2.20)	-.4747 (-1.28)	-.2007 (-0.90)	N.O. —	-.0276 (-0.15)	-.6015 (-2.22)
San Luis Potosí	-.0709 (-1.29)	.156 (0.06)	-.519 (-0.33)	-.3938 (-0.34)	-.1212 (-0.72)	-.0647 (-0.32)
Tampico	0.555 (0.61)	.1157 (0.45)	.1128 (0.21)	.2742 (0.24)	.7119 (2.86)	.0462 (0.26)
Reynosa	.1973 (3.34)	-.0771 (-0.30)	.4952 (2.23)	.6386 (0.56)	.3357 (1.48)	.4993 (3.32)
Mexicali	.0080 (0.11)	.0850 (0.30)	.0060 (0.03)	.4660 (0.32)	.3128 (1.34)	-.0449 (-.013)

Table A.6. (*continued*)

	White-Collar		Blue-Collar		Tertiary	
	Registered (N = 1,032)	Unregistered (N = 105)	Registered (N = 152)	Unregistered (N = 47)	Registered (N = 210)	Unregistered (N = 444)
Mazatlán	.1939 (3.46)	.3835 (0.74)	.1010 (0.19)	.7866 (0.44)	.4620 (1.45)	.1074 (0.43)
Oaxaca	-.2187 (-4.09)	-.7803 (-2.40)	-.3623 (-0.70)	-.0858 (-0.08)	.3099 (-1.62)	-.5527 (-4.45)
Villahermosa	-.0032 (-.040)	N.O. —	-.5214 (-1.15)	N.O. —	-.2007 (-0.70)	.3044 (0.81)
Mérida	-.2885 (-4.30)	-.2091 (-0.74)	-.6203 (-2.15)	-.3478 (-0.27)	-.0634 (-0.33)	-.3474 (-1.79)
R^2	.230	.241	.297	.214	.056	.386

Note: The dependent variable is the natural log of real peso earnings. The constant term, or base, represents Venustiano Carranza and workers with primary education incomplete and over one year on their current jobs. N.O. denotes variables for which there were no observations.

Numbers are regression coefficients; *t*-values are in parentheses.

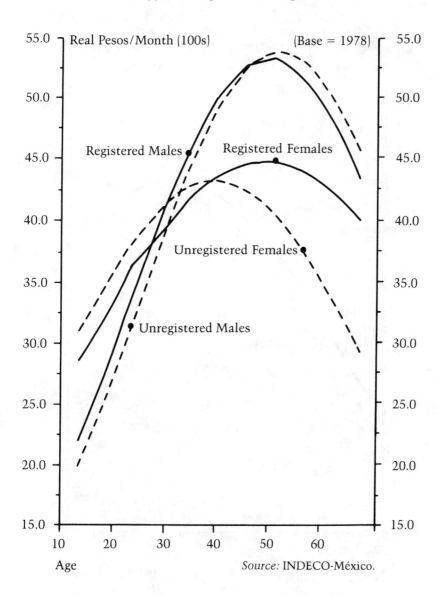

Figure A.1 **Age-Earnings Profiles: Registered versus Unregistered White-Collar (Primary Education Only)**

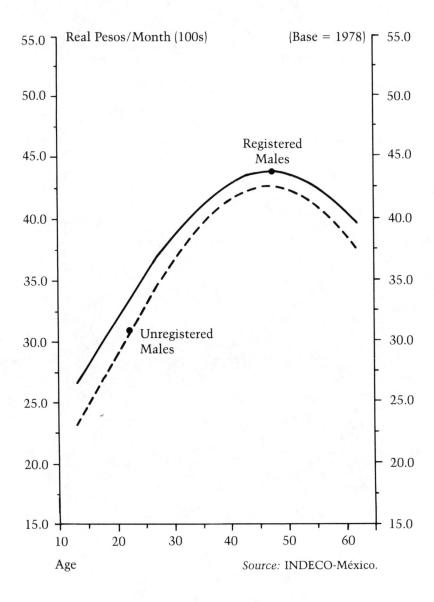

Figure A.2 **Age-Earnings Profiles: Registered versus Unregistered Blue-Collar (Primary Education Only)**

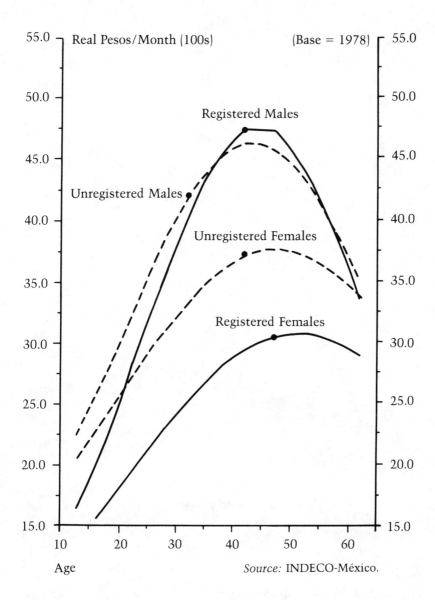

Figure A.3 **Age-Earnings Profiles: Tertiary Registered versus Tertiary Unregistered (Primary Education Only)**

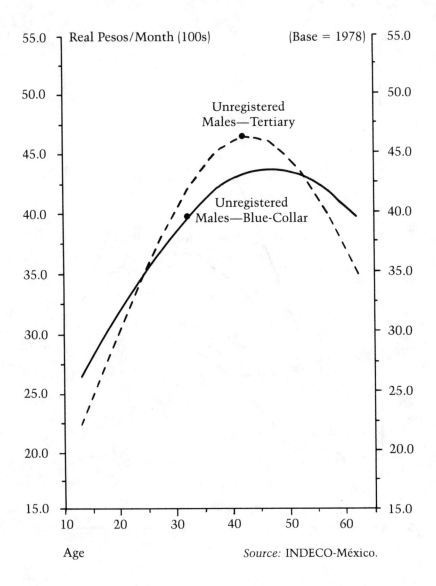

Figure A.4 **Age-Earnings Profiles: Blue-Collar Registered versus Tertiary Unregistered (Primary Education Only)**

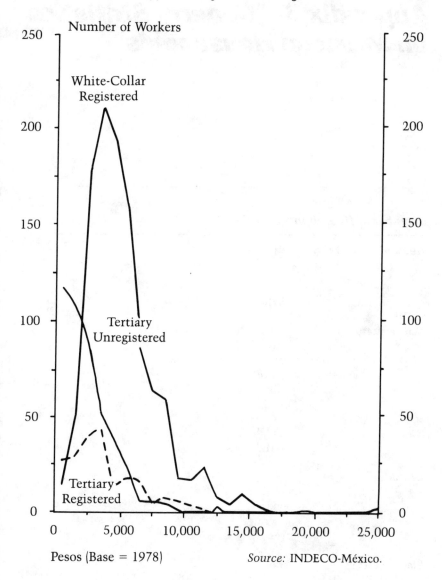

Figure A.5 **Distribution of Real Monthly Earnings: Females (Primary Education Only)**

Appendix 3 General Statistics on Mexican Households

(Number of Households Sampled 9,459)

Head of household

Age

	Mean	41
	Median	40

Marital status

	Married	85%
	Divorced/separated	2%
	Widowed	6%
	Single	7%

Sex

	Female	12%
	Male	88%

Education

	Primary or less	63%
	Secondary	16%
	Superior	12%
	University	10%

Place of birth

	The city	40%
	Same state, rural	17%
	Other city	8%
	Other state, rural	14%
	Other state, urban	21%
	Foreign	1%

Occupation

Agriculture	5%
Casual labor	8%
Low-level services	17%
Blue-collar	19%
White-collar	40%
Business/professional	10%

Job stability

Stable	71%
Unstable	29%

Work sector

Unemployed	3%
Formal	59%
Informal	33%
Patron	4%

Income

Mean	$5,973
Median	$4,130

Monthly income in minimum salaries

<1	25%
1–2	48%
2+	27%
Mean	2.13
Median	1.28

Fringe benefits

None	45%
Medical	14%
Medical + other	30%
Miscellaneous	11%

Work sector

Formal	55%
Informal	32%
Student/housewife/unemployed	5%
Unknown	8%

Percentage in labor force

Not in labor force	13%
In labor force	87%

Characteristics of dwelling

House type

Shack	7%
Apartment	12%
Room	9%
House	73%

Uses of house

Residence only	94%
Commercial/business	6%

Floor area (M^2)

Less than 50	29%
50–99	36%
100–149	18%
150+	17%
Mean floor area	147M^2
Median	78M^2

Number of floors

1	87%
2	12%
3+	1%

Rooms in dwelling

1	16%
2	23%
3	21%
4	18%
5+	21%
Mean	3.2
Median	3.0

Household demography

Where household formed

This city	65%
Same state, urban	8%
Same state, rural	8%
Other state, urban	13%
Other state, rural	5%
Foreign	1%

Reason for migration

Work	65%
Home ownership	5%
Education	7%
Family	13%
Other	11%

Date of migration to city

Median	1966

Date of arrival, current location

Median	1970–1971

Migrants from household

No	67%
Yes	33%

Number urban-to-rural migrants

0	89%
1+	11%

Number urban-to-urban migrants

0	77%
1+	23%

Number migrants to Mexico City

1+	7%

Migrants to foreign countries

1+	11%

Number of dependents under 16 or over 65

Mean	3.6
Median	3.2

Number economically active (aged 15–64)

Mean	3.0
Median	2.4

Age dependency ratio

Mean	1.5
Median	1.0

Parent in household

	Yes	9%
	No	91%

Number of children in household

	None	17%
	1	16%
	2	16%
	3	15%
	4	12%
	5	9%
	6	6%
	7	4%
	8	5%
	9+	0%
	Mean	2.9
	Median	2.5

Siblings present in household

	None	95%
	1	3%
	2	1%
	3+	1%
	Mean	.1
	Median	.03

Grandparents in household

	None	95%
	1	4%
	2	1%
	3+	0%

Grandchild in household

	None	98%
	Yes	2%

Other relative in household

	No	95%
	Yes	5%

Family type

	Singleton	2%
	Matrifocal	7%
	Nuclear	73%
	Complex	17%

Number of income producers in household

None	5%
1	65%
2	20%
3	7%
4	2%
5	1%
6+	1%
Mean	1.4
Median	1.2

Number of household members not in work force

Mean	3.9
Median	3.6

Worker dependency ratio

Mean	3.4
Median	3.0

Number of people under 21 in work force

None	86%
1	11%
2	3%
3+	1%

Stage of life cycle

Young couple, no children	10%
All children under 5	36%
All children under 15	31%
All children over 15	15%
Older couple, no children	7%

Total family size

0–4	35%
5–7	36%
8+	29%
Mean	8.2
Median	5.5

Participation in neighborhood organizations

No	89%
Yes	11%

Household economics

Total household monthly income
Mean	$13,994	(US$615, approximately)
Median	$5,034	(US$221, approximately)

Monthly household expenditures

Household upkeep costs (not rent)
None	59%	
Mean	$492	(US$21.50)

Taxes
None	52%	
Mean	$165	(US$7.25)
Median	$0	

Food
Mean	$2,773	(US$122)
Median	$2,500	(US$110)

Water
None	17%	
Mean	$42	(US$2)
Median	$40	(US$2)

Fuel
None	1%	
Mean	$147	(US$6.50)
Median	$80	(US$3.50)

Electric
None	9%	
Mean	$222	(US$9)
Median	$120	(US$5)

Transportation
None	14%	
Mean	$389	(US$17)
Median	$200	(US$7)

Medical
None	58%	
Mean	$152	(US$6.50)
Median	$0	

Entertainment

	None	55%	
	Mean	$243	(US$10.50)

Installment payment

	None	62%	
	Mean	$249	(US$10.50)

Savings

	None	75%	
	Mean	$221	(US$10)

Other

	None	69%	
	Mean	$275	(US$12)

Fuel used in household

Piped gas	4%
Bottled gas	84%
Kerosene	8%
Wood	2%
Coal	1%

Transportation to work

Bus	47%
Taxi	3%
Car	29%
Foot	15%
Bicycle	4%
Motorcycle	1%
Animal	0%

Lot tenancy

Legal ownership	48%
Irregular ownership	21%
Regular rental	23%
Irregular rental	4%
Borrowed	4%

House tenancy

Regular ownership	54%
Irregular ownership	16%
Regular rental	23%
Irregular rental	3%
Borrowed	4%

Quality judgments

	House	
Very bad	2%
Bad	5%
Medium	24%
Good	45%
Very good	24%

Household income in minimum salaries

<1	23%
1–1.99	38%
2–3.99	25%
4+	14%
Mean	4.2
Median	1.65

Work sector of family members

Formal	54%
Informal	29%
Mixed	7%
Student	1%
Unknown	9%

Household income spent on food

0–19	14%
20–39	29%
40–59	29%
60–79	19%
80+	9%
Mean	44.9%
Median	42.9%

Obligatory expenses

Mean	$3,542	(US$156)
Median	$3,252	(US$143)

What house has

Electricity	88%
Sewer	62%
No water	19%
Water in patio	60%
Water in house	21%
Separate kitchen	95%
Bathing facilities in house	56%

Notes

1. The Setting and the Study

1. A pseudonym, as are the names of all people in the book.
2. Throughout this book we use the term "ordinary" to describe the majority of families in Mexico. We use it with the same tone of respect as Lincoln reserved for the "common man" which, as everyone knows, God loved greatly otherwise He would not have made so many of them. We use the term "ordinary" for two reasons. First, we do not wish to imply that we carried out a proper class analysis. This analysis is being done in Mexico, in two traditions: (1) a Marxist tradition emphasizing the relationship of householders to the means of production, as in the work of García, Muñoz, and Oliveira (1982) for Mexico City and in the work of the "Jalisco group," as for example, in the work of de la Peña (1986), Escobar (1986), González de la Rocha (1986), Gabayet (1988), and Lailson (1985) (see the Bibliography for citations to recent work). In a more descriptive and concrete vein, we have the work of Lewis (1961), for example, on the working class or more recently Ramírez (1979) on the culture of the working class, as well as that of Nutini and Roberts (n.d.) on the aristocracy, and Lomnitz (1987) on the entrepreneurial class, as well as an emerging literature on the middle class. A thorough treatment of the Mexican class system has not been done.

The second reason we use the admittedly awkward term "ordinary" to describe the majority of Mexican householders is that we want to avoid the connotations that terms like "working class" and "middle class" have for American and European readers. Some writers try to avoid the theoretical difficulties by using another terminological trick, describing households that are neither poor nor middle-class-in-the-Euro-American sense as "middle sectors." We sometimes use the term "middle and upper sectors" to refer to the top 10 percent of our sample. We are not suggesting the term "middle sectors" because our middle and upper sectors only represent 10 percent of the Mexican population. Our sampling strategy underestimated the dominant sectors in the society, since our sample was confined to the provinces and a working-class area of Mexico City, where there are fewer middle- and upper-class families than in the capital city's posher residential areas. In any case we use income criteria to stratify the urban population. Our "ordinary Mexican households" belong to the great mass of people whose total household income (all sources) is less than 4.8 minimum salaries. This amount, which varied from a low of Mex$8,640 per month in Oaxaca and Reynosa (or US$380 in 1980 dollars) to a high of Mex$17,280 (or US$794.50 in the Federal District and in Mexicali) is what used to be called the FOVI standard (FOVI meaning Fondo de Vivienda, or Fund for Housing). Households earning between 1.8 and 4.8 minimum salaries qualified for mortgage money from government subsidized funds. Below 1.8 they were too poor to qualify, and above 4.8 they were deemed fundable in the private mortgage market. Over 90 percent of our sample had household incomes less than or equal to 4.8 minimum salaries. We break these households down into two categories: "the poor," households that earn 1.8 minimum salaries or less, which we consider to be living below the poverty line and which make up around 50 percent of our sample; and "those who are getting by," or households with more than 1.8 minimum salaries and less than 4.8, which make up around 40 percent of the sample. More details on income are given in chapters 4–6.

3. *Educación superior,* or postsecondary education, refers to formal education past the ninth year either in a technical school or a preparatory school.

4. "Ordinary Mexicans" have always been a majority of the population. Their numbers and their fraction have increased as a result of the economic crisis that has been going full force since 1982.

5. The worker dependency ratio is the ratio of the number of members of the household not in the work force to the number in the work force. The average number of household members in the work force is 1.40, while the average number of members not in the work force is 3.98, yielding an overall dependency ratio of 2.8.

6. A "parachutist" is a *paracaidista,* a person, family, or household that is an urban squatter who has just invaded some area.

7. Murphy and Stepick are preparing the manuscript for a book titled "Adaptation and Inequality in Oaxaca: Political Economy and Cultural Ecology in an Intermediate Mexican City."

8. In the 10-year follow-up study in Oaxaca, 1987, we asked people about their plans to move, and the answers were interesting. Those who had considerable investments in their houses indicated little inclination to move. Women were less inclined to have any desire to move than men (30% vs. 50%). The men who were most inclined to move were those who had a history of moving in the past. There was no indication that an American middle-class rationale for "moving up" or treating one's house and site as a liquid investment was to be found among the ordinary families in Oaxaca.

9. This is a shaky statistic, really meant to give the reader some idea of the extent of popular housing in Mexico. Although *colonias populares* are

not so extensive in Mexico City proper, the figure is roughly accurate for the metropolitan zone as a whole.

10. As noted later in this chapter, we have followed the Fund for Housing in establishing the poverty line at a household income of 1.8 minimum salaries. One minimum salary is not nearly sufficient to maintain a family at minimum levels of living. Such families would show nutritional deficiencies and be unable to participate at the culturally appropriate level in community activities. For lack of a better term, we might call these the "immiserated poor."

11. A rather upscale term for a downscale technology; milpa agriculture with a *yunta* technology means raising corn, beans, and squash in mainly unirrigated fields with a yoke of oxen as tractor power.

12. Jesús Sánchez still lived there, although he had bought a house in Ixtapalapa with his share of the royalties of Lewis' enterprises. He still worked three times a week at the restaurant where he had worked during the period covered by Lewis.

13. Middle-class people tend to take privacy for granted, since we have enough space to defend ourselves. For poor people in village or city, it is sought after. One of us invited his *compadre* from the village to stay in Austin for a couple of months and subsequently asked him in the village what he had found to be the most interesting thing about living in the house in the States. The reply was interesting, "You can turn on the music, smoke and drink and dance in the house, and no one, not even the neighbors, knows what is going on." Later, when we tried to find accommodation in Ciudad Netzahualcoyotl, which is very full and not at all designed for rental properties, we tried to rent a place that lacked a *barda*. Our overly generous offer was refused, however. The owners saw all the project equipment and assured us that without a *barda* we could never defend ourselves and that the equipment would be gone within a week.

14. It is not a coincidence that there is no Mexican translation for this North American archetype. The common man is a theme that runs through the literature on *lo mexicano* and has been discussed in its forms as the *pelado* (Samuel Ramos, 1934) and the *chingado* (Octavio Paz, 1961), but the only time one encounters Lincolnesque types among the ordinary people in Mexico is every day in every *colonia popular* and the requisite vocabulary of praise for this type is lacking in Mexican arts and literature.

2. Cities of the Study

1. These terms will be defined momentarily when they are integrated into the discussion. The terms refer to the potential of a city's industrial base to create good jobs.

2. One has to be careful here, since there are regional differences as well as class differences. In the metropolitan zone of Mexico City, ordinary people do tend to a more "middle-class" life style than they do in the

provinces but, at the same time, not so much as to disable the generalizations about the importance of the family as the source of leisure, entertainment, and emotional restoration.

3. It is improper to speak of these villages as part of the "reserve army" of laborers as some commentators have. A reserve army has to be a group that is disciplined for industrial work, trained and ready in reserve, ready for the cyclical upturn that will call them to the colors. Rather than regarding the villages as the bivouac area of the reserve army, it is more accurate to think of them as warehouses where between five and 10 million rural Mexicans are housed at minimum cost to the state and to the system of production.

4. These figures are for the cities as recorded in the 1980 census and do not refer to metropolitan areas. The metropolitan areas are usually two to three times the size of the cities themselves. The Census Bureau is in the process of creating standard urban census areas, but the figures for these areas were not available at the time of writing.

5. Not true of the middle and upper classes, who do use it in the normal European fashion as a pejorative. This neutral use of the term contrasts with the plethora of terms that Mexicans of every region use to contrast themselves with other regions. For people in the north, residents of Mexico City are *guachos*, while the Yucatecans use the same term for people who are not Yucatecos. Many regions refer to residents of Mexico City as *chilangos*, who are known to speak *caló*, the dialect of El Tepito in Mexico City, which *chilangos* refer to as *caliche*, using the *caló* word for *caló*. *Chilangos* are especially apt to use the term *naco*, which most likely is derived from the Otomí word *naco*, which means brother-in-law, but which in popular etymologies is thought to derive from *totonaco*, which means "truly Mexican" in the long form and "sucker" or "bumpkin" in the short. There is no shortage of regional typifications in local speech, much more so than in American English.

6. Lorenzen (1986) has shown that the only group that can be said to belong to an "aristocracy of labor" are the *petroleros*. They are the only workers for whom the unioned formal sector status yields clear advantages in salary and standard of living.

7. We cast about for a long time to find the best available measure of good labor market conditions and learned a great deal about the dearth and unreliability of statistics on labor markets and economic indices. To give but one example, we had thought that the incremental capital-output ratio (Scott, 1982:99), so enthusiastically endorsed as a development index in the second development decade, might well serve as an indirect measure of the growth potential of a city, which was what we felt ordinary people were concerned about when they evaluated their prospects in the city. But the data that make up this index are of dubious value and/or very difficult to evaluate. The lowest capital-output ratios are found in Villahermosa (3.7), Querétaro (4.1), Mérida (4.3), and Reynosa (5.1), while the highest are found in Oaxaca (33.6) and Mexicali

(30.4). It did not seem to us that there was any a priori homogeneity to the categories, so that, in the latter case, Oaxaca, a sleepy, low-wage, colonial, nonindustrial, provincial town of tourists and bureaucrats could be grouped with a rapidly expanding frontier town full of immigrants and aspirant U.S. residents, transport workers, unions, and some of the most productive agribusiness in the country. Second, the expected kind of correlations between the capital/output ratio and average plant investment, plant size, and the like are as near as you like to zero, which suggests the dubious applicability of the index.

8. Though perhaps a little out of place in a book of this kind, one might mention the best recent work on the influence and politics of the national oil company and its union, as well as a description of the union's many corporations and their activities and the many advantages that accrue to workers for belonging. It is a novel, *Morir en el Golfo*, by Héctor Aguilar Camín (1986). In addition, much has happened since February 1989, when the newly elected president of the nation arranged for a dramatic entry into the home of the leader of the *petroleros* in Ciudad Madero, where arms were found and thus grounds for the charges on which he was jailed. The oil union is still powerful and rich, but it pays a good deal more attention to the central government than it had been accustomed to in the past.

9. A list of possible reasons was given to respondents, and they were asked to give the most important one insofar as it affected their own decision to come to the city in which they were now living. The overwhelming response had to do with jobs (62%). Family was in second place (12%), while education and home ownership were given by 6 percent of the respondents each. Twelve percent gave miscellaneous reasons.

10. Unless otherwise mentioned, all peso figures are in real Mexican pesos, with a base equal to the Federal District, 1978.

11. Technically, we should not be using the term "families," since our major data source was based on households, but as can readily be imagined, households are based on families and the cultural codes for household, family, and kinship show a great deal of interpenetration and overlap.

3. Households, Strategies, and the Economic System

1. These studies may be out of date. On the basis of an admittedly slender reed, namely our interviews and survey of the city of Oaxaca carried out in 1987, our very definite impression was that, because of the seemingly permanent economic crisis, the networks of exchange that had been characteristic of households during the epoch of the great rural-to-urban migrations (1950—1970) had degraded considerably and that now the urban family is very isolated. Investigations on this important topic proceed.

2. *New York Times*, January 7, 1988. The *Times* noted that the proportion

of black matrifocal households had increased 160 percent since 1968, while the proportion of white matrifocal households had increased 288 percent in the same period.

3. Perhaps too fussily, we distinguish between gender concept, which refers to the ideas that the people we are talking about have about what they consider male and female, and "gender construct," which is the analytic term that we use for comparative work. They are both different from sex role.

4. This account is mostly based on observation and living in generous, long-suffering Mexican families who have housed and suffered the investigators over the years. Some of the most interesting data was collected at night while watching *telenovelas* with ordinary Mexican families. As aficionados of the genre know, family and kinship relations can become quite bizarrely involuted in these television dramas, and it often required a good deal of explanation about how households, families, and kin groups ought to work in order to explain the plot to the profane visitor.

5. We thank Orlandina de Oliveira for this suggestion, given in a course of four seminars on the household in urban Mexico at the University of Texas at Austin in October 1987.

6. This brotherly obligation is not nearly so strong as it was in the 1950s if we can judge from the nearly obsessive preoccupation of the Sánchez boys with their sisters' virginity and their endorsement of the double standard in sexual expression. But Lewis is perhaps suspect in his emphasis here, influenced as he was by psychoanalytic preoccupations.

7. Hugo Nutini, who has spent a lifetime studying the communities of Tlaxcala and whose knowledge of the area is unequaled, mentioned this in conversation.

8. It has been estimated that in the 1970s between 10 and 12 percent of the households, to a total of 25,000, were involved in *maquila* work in Ciudad Netzahualcoyotl.

9. González de la Rocha's descriptions of Chabela, who sewed peaks for workmen's peaked caps, is as good as Lewis' descriptions at their best. On "good" days, Chabela would sew 700 peaks and hand them over to her *patrón* the next day, when he would supply her with more materials. Meals consisted of tortillas and beans, for that was all she had time to cook. On busy days, she was up at 5:00 or 6:00 in the morning in order to have some time to cook something and to get the house in order. She started sewing around 8:00 in the morning and kept at it until 3:00 or 3:30 in the afternoon. Sometimes she had to sew right on into the night because she had to hand over the peaks to her *patrón* the next morning. Sundays were her only day off, and they were spent almost entirely in straightening up the house and getting the clothes ready for herself and her children for the week to come. Still, she preferred it to factory work, which many women do in Guadalajara, because it meant that she could be with her children.

10. Rosenzweig (1973, 1977) is an exemplary study in this tradition, which links children's quantity with both their costs and their value in farm work. The first deals with the supply of rural children historically in the nineteenth- and twentieth-century United States, and the second is a cross-cultural study of the supply of children and their value in rural India. To us the fit he gets between his predictive equations and the empirical rates is very impressive.

11. Once again our data is drawn from the follow-up study in Oaxaca, where we find that although the correlation between the economic status of the household and the number of children in it has remained in effect in the postcrisis period (i.e., richer households have more children), the absolute number of children and the number of children with earnings have declined. This relationship will be examined in chapter 5 for the precrisis period and in chapter 9, where the comparison between the precrisis and the critical period is brought out.

12. To our knowledge no similar study has been carried out for Mexico.

13. Later in his life, when he was in his late seventies, Sánchez was still lamenting his failure, as he confessed to Selby, who interviewed him in 1985. He was very grateful to Lewis, who had made him famous, which he liked a lot, and who had paid him deservedly well, so much so that he had built a very nice house in the suburbs, and his children had been able to educate themselves more fully and all get jobs either in the bureaucracy or in the professions and join the lower middle class. But Sánchez claimed that Lewis had exaggerated the bitterness of his feelings about the child called Robert and had also exaggerated by far the degree to which Robert had been a bad boy or had disappointed his father. He was still lamenting the fact, he said, that his children did not visit him, and although he realized that they had separate careers and lives, he still harbored a wish that one of them would come to live with him in his suburban house.

We did not know quite what to make of this testimony. The difference between their lives as depicted in *The Children of Sánchez* and in 1985 was so dramatic and striking that one was persuaded that changed circumstances had changed memories of how difficult it had been, and how difficult a father he had been, during this very trying period of his life.

14. We would include our own work which uses linear programming to try empirically to derive the implicit utility function under which people are operating as one of the failures, in this sense. (For urban populations, see Murphy, 1973, and Stepick, 1973, and for rural populations, see Selby and Hendricks, 1976. The problem in our case lay with defining and estimating the equations that linked activities to other activities, to resources, and to goals. Theoretically the problem is solvable, in that multidimensional scaling techniques can be utilized to define real numbered scale points that can be correlated with the different levels of whatever goal is being pursued. For example, one can scale an activity

like "activity in a cargo system" and correlate it with the values of a variable such as "respect," and one can compare optimal with observed activity levels, as we tried to do. But there is no evidence that the people whom we studied make judgments that correspond with our measures, or even that they could even if they wanted to. In a word, a computer will change the trajectory of a behavioral system when it encounters a critical value or threshold of some kind. People cannot make and do not make precise estimates of critical values.

15. Official statistics on the work force show that 71.1 percent of the men and 16.4 percent of the women 12 years old and older are in the work force (SPP, 1982:60). Women's participation rates vary by region, being the highest in the metropolitan zone, and among women ages 20–24 (24.1% of the work force in 1970 [ibid.]). According to our data collected in 1976–1978, 78 percent of the work force is male, and 73 percent were heads of households. Since the median age of the household heads is 41, the typical citizen that we observed going home was a middle-aged male worker, who was the head of a household.

16. Urban unemployment is high but not nearly as bad as unemployment in the agrarian labor force, which is estimated by the Economic Intelligence Unit's *Mexico Country Report* (1988) to reach 47.6 percent by the end of 1988.

17. In this book "the state" is not theorized. It is shorthand for the external forces that impinge upon the household either consciously or analytically. Doug Uzzell, who has done some of the most interesting work on the relationship between the household and the state in Latin America (Uzzell, 1974, 1980, 1987), has warned us that to talk of the state as a whole rather than the particular institutions, agencies, ideological moments, and concrete points of articulation between the household and the state (as he has done in his work in Lima, Peru) is to leave out an important and interesting part of the analysis. We agree. But we have not done the field work, and so our "state" is a shadowy, empty affair only because it has to be.

18. Terms as defined in de Janvry (1981): "market widening" is the process of incorporating more consumers into the market; "market deepening" is the process whereby consumers gain the capacity to buy a greater variety of more desirable consumer goods. Market widening is associated with the process of proletarianization of the peasantry; market deepening, with the creation of a middle class and an increase in equity in income distribution.

19. This presumably will be changing under the current process of entry into GATT, which is currently projected to take place over the next 10 years. We shall discuss the effects of this process, as well as the effects of the changes being brought about by the currently five-year-old economic crisis, in the final chapter, under the heading of the "New Deal" for the Mexican workers.

20. This statement requires another book of justification, and the argument will not be further pursued here. Anyone who has taught or observed in

the system of public education in Mexico cannot fail to be impressed by the omnipresence of authoritarian relations and the promotion of an authoritarian view of Mexican society. The emphasis on *compañerismo* (comradeship) in left-wing "free schools" and/or voluntary schools, which British or U.S. citizens would find false and self-conscious, is a directed attempt to undermine and oppose the authoritarian rule and message of the school system. In smaller, especially rural schools, the arrogance of the teachers is breathtaking, as they denigrate the children at the same time as they lead them in mindless chanting of written materials in place of learning. Capriciousness is built into the system, most important, in the secondary admissions process and in admission to subsequent levels of education, especially *educación superior*, where a lottery is used to choose among the qualified. It would be simple to raise the grade standard so as to choose students on meritocratic grounds, but instead a lottery, with all the opportunity it provides for side payments to the authority is preferred. The mystification of *lo mexicano* is a constant theme, particularly in the primary years where lesson materials emphasize the uniqueness of the Mexican experiment and its fundamental *mestizaje* (i.e., lack of racism) in the creation of a fair society in which there are no upper-class *güeros* (fair-skinned people) who run things and no marginal *indígenas* who barely survive as the most exploited group of the population.

4. The Mexican Urban Household

1. The household, a residential group, is based on a family type. "Nuclear household" means "a household based on and containing a nuclear family."
2. Hotchkiss (1968) has discussed the important roles of children in Mexican villages. Living without children in the villages and cities of Mexico, as we have, brings the lesson home. One is compelled to "borrow" them.
3. The work of Kim (1988) on female labor force participation in urban Mexico, based on the same data being analyzed here, shows that the female labor force is split in two. A high-status, well-educated, mainly young and unmarried group suffers from little if any wage discrimination. An older group of late entrants or reentrants into the labor force suffers more, earning 60 percent of what men earn. Our earnings reports suffer from lack of data on hours worked.

5. Household Dynamics and Economic Survival

1. Those earning over 4.8 minimum salaries would consider themselves middle class (or upper class), and on the grounds of consumption styles and occupation, Europeans and Americans would tend to concede them that status. In our sample they are only the top decile of the household income distribution. We undersampled in the metropolitan zone of the capital, where around 40 percent of the country's population and a ma-

jority of its middle and upper classes live, so they are underrepresented in our sample. Our metropolitan sample was carried out in the *delegación* of Venustiano Carranza, which is a working-class area stretching from the eastern part of La Merced to the area around the airport.

2. The major item is food which takes up about 41 percent of the average budget.

3. Or were at the time of the 1977–1979 survey. The administration of President Miguel de la Madrid (1982–1988) reduced the subsidies for food greatly.

4. The rock bottom expenses for secondary school in Oaxaca in 1987 were reported to be between US$5 and 7 for enrollment, and an equal amount in "contributions" which children themselves are expected to raise through the selling of raffle tickets (*boletos*) which the majority of parents end up buying. Book expenses can vary, but they rarely exceed US$30 a year. Uniforms are expensive, but they can sometimes be avoided by sending one's child to a school that has a "uniform-free" session (usually the afternoon).

5. This is changing rapidly right now, and the changes in the cost-benefit equation that influences the value of children will be discussed in the last chapter.

6. Workers, Jobs, and Salaries in Urban Mexico

1. We apologize to readers who are offended by the term "workers" in reference to "labor force participants." This whole book insists that the work of social reproduction, household maintenance, and psychological endurance is based upon the unremunerated work of men, women, and children in the household, particularly the latter two. So we are not suggesting that people who are not in the labor force are not working because, of course, they are. But "labor force participation" is awkward stylistically, so we beg indulgence.

2. See Leeds (1977) for a discussion of the concepts regarding urban employment, which has been very helpful to us.

3. A detailed analysis of female labor force participation, based on the INDECO data set, is Kim (1987).

4. See Kim (1987).

5. Overall, women have slightly fewer years of schooling than men. But the situation is reversed for people in the labor force: women have 8.1 years of education, and men have 6.1 years.

6. Kim's analysis shows that the younger women in white-collar jobs earn about 90 percent of the wage of male workers, while the older women who tend more to be engaged in casual, sales, or other low-level service jobs earn about 60 percent of the male wage. (For details, see her work and our Figure 4.1 and associated tables.)

7. "Job stability" is evaluated from a survey question that asked all people in the work force whether they had held their current job for a year or more ("stable") or "less than a year" ("unstable").

8. For the reader unfamiliar with the term "dummy," a dummy, or a dummy variable is a binary valued variable (o, 1) that is defined in a regression equation to represent the absence or presence of some characteristic.

9. The professional-entrepreneurial group is of relatively small consequence as a viable employment option for the great majority of the urban workers that are the focus of this study. This group is therefore not of primary interest in the present study and is consequently not included in the statistical analysis that follows.

7. Household Income and Economic Welfare

1. We hope to be forgiven this "Spanglish" expression that crops up frequently in philosophical discussions of household and family among ordinary people in Latin America. Some social scientists may regard households as maximizing firms, or as the conceptual units of social reproduction, but to ordinary people in urban Mexico they are the collective means for "defending themselves." Incomes are also seen this way, not as the means of acquiring some set of material goods. With money you can defend yourself. The term itself, as can be imagined, means everything. It ranges from the individualistic meaning of not allowing yourself to be vulnerable to others, *no rajándose* ("keeping on," "bearing up," "making yourself less vulnerable") as in the famous discussion in Paz' (1953) *Labyrinth of Solitude,* to the notion of "knowing how to make it" as a family and kin unit or, in our phrase, "getting by."

2. These have been called "obligatory expenses" or just "expenses" in chapters 4 and 5.

3. It is corroborated by studies of the household in other parts of Mexico, however, as González de la Rocha's (1987) studies of the Guadalajara region have shown.

4. The differences between men and women were insignificant: 37 percent of the men and 39 percent of the women endorsed the use of birth control and the idea that one could not be happy without children.

5. The figures are quite decisive. In the domestic cycle, 42 percent of the second quintile are in the stage where all the children are under five years of age, while only 29 percent of the fifth quintile are in this stage. The dependency ratio for the second quintile is 3.5 (dependents per member in work force), while for the fifth it is 2.4. The median age of the household head is 40 for the second quintile and 43 for the fifth.

8. Household Budgeting Strategies

1. Musgrove (1978) has shown convincingly that consumption-based estimates of permanent income are greatly preferable to income-based estimates because of the variation in incomes as a result of sporadic employment and underemployment, which forces householders to change

jobs frequently and to look for income-earning opportunities where they can.

2. The *prestaciones* that we specifically interviewed for were (1) social security (including and mainly medical and hospital benefits; (2) "social benefits" like vacations; and (3) economic benefits like end-of-year bonus (*aguinaldo*), sick pay, subsidized housing opportunities, and pension benefits.

3. One cannot help but remark on the contrast between this voluntary massive movement to a new housing area with the plethora of studies on new towns and new settlements that began with *Family and Kinship in East London,* which showed how reluctant people were to take up residence in new areas because of the lack of social bonds, splitting of families, lack of amenities like the old pub, and the destruction of commercial opportunities that had abounded in the old neighborhood. There are few complaints about having to move out of the old neighborhood here, even though it was and still is the commercial center of working-class Mexico City and contains one of the finest markets for contraband in the world. Everyone expressed relief at being able to get away from the rents and at having the chance to spread out a little after the cramped and costly quarters of the old neighborhood.

4. One of us carried out an intellectual exercise while doing field work in Ciudad Netzahualcoyotl, conducting an informal, outdoor, Sunday morning "seminar" among his neighbors. There were two texts for discussion: *Los Hijos de Sánchez* by Oscar Lewis, which many of them knew not because they had read the book (only one seminar member had) but because they had lived in the neighborhood, and *Worlds of Pain,* by Lillian Rubin, which is a study by a psychologist of the difficult world of the working class living just above the poverty line in the San Francisco Bay area in the early 1970s. The Mexicans thought, not unnaturally, that the Americans hardly had anything to complain about, since their living standards were so much higher than the street seminar's, but once induced to be culturally relative, they agreed that the big difference between the two lives was the "bills" that the Americans worried about all the time and that seemed unavoidable in their lives. The seminar members all agreed that one did everything in one's power to prevent the accrual of bills, payments that had to be made every month whether there was income or not. They agreed that they would be equally worried, but had to wonder about why the Americans did not avoid them.

9. The Economic Crisis and the New Adaptation

1. We would like to take this opportunity to thank Laura Finsten, professor of anthropology at Macmaster University, Hamilton, Ontario, for her staunch help and companionship during the interviews, which would not have been as possible nor as pleasant without her company and contribution.

2. Don Pedro is a brand of good quality brandy and often is bought in quarter liter bottles. The term is sometimes used to refer to a policeman's bribe, as in the expression, "Me hace falta mi Don Pedro."

Appendix 1. Study Methods

1. The MIT method relies heavily on air photos, but in Oaxaca the latest photos of the city dated from 1973, three to four years before the study period and were considerably out of date. The project did take air photos, but they were not of sufficiently high quality for the detailed typological work envisioned by the MIT method. Walking the city proved much more satisfactory and gave the project members a much better idea of the city.
2. Within each locality segment, a primarily residential block is selected to allow comparison of areas and densities that are homogeneous. The block is bounded on all sides by circulation, so that the ratio of circulation (or service) to areas can be compared.

Bibliography

Alonso, Jorge, ed.
  1980   *Lucha urbana y acumulación de capital.* Mexico City: Casa Chata.
Alonso, José Antonio
  1981   *Sexo, trabajo y marginalidad urbana.* Estudios Sociales. Mexico City: Edicol.
Argüello, Omar
  1981   "Estrategias de supervivencia: Un concepto en busca de su contenido." *Demografía y Economía* 15(2):190–203.
Arias, Patricia
  1982   "Consumo y cooperación doméstica en los sectores populares de Guadalajara, Jalisco." *Nueva Antropologia* 6(19).
Arizpe, Lourdes
  1982   "Relay Migration and the Survival of Peasant Households." In *Towards a Political Economy of Urbanization in Third World Countries,* ed. Helen Safa. Mexico City: Secretaría de Educación Pública.
Auletta, Ken
  1983   *The Underclass.* New York: Vintage.
Balán, Jorge, Harley L. Browning, and Elizabeth Jelin
  1973   *Men in a Developing Society: Geographic and Social Mobility in Monterrey, Mexico.* Austin: University of Texas Press.
Baldwin, John
  1974   *Guide for the Survey-Evaluation of Urban Dwelling Environments.* Cambridge: MIT Press.
Becker, Gary S.
  1964   *Human Capital: A Theoretical and Empirical Analysis.* New York: Columbia University Press.
  1976   *The Economic Approach to Human Behavior.* Chicago: University of Chicago Press.
  1981   *A Treatise on the Family.* Cambridge: Harvard University Press.
  1986   "Family." In *The New Palgrave: A Dictionary of Economics,* vol. 2, ed. John Eatwell, Murray Milgate, and Peter Newman. New York: Macmillan.

Benería, Lourdes
   1979   "Reproduction, Production, and the Sexual Division of Labor."
      Cambridge Journal of Economics 3(3):203–225.
ben Porath, Yoram
   1982   "Economics and the Family: Match or Mismatch." *Journal of Eco-
      nomic Literature* 20: 52–64.
Berger, Suzanne, and Michael J. Piore
   1980   *Dualism and Discontinuity in Industrial Societies.* Cambridge:
      Cambridge University Press.
Bowles, Samuel, and Herbert Gintis
   1976   *Schooling in Capitalist America.* New York: Basic Books.
Buchler, Ira R., and Henry A. Sclby
   1968   *Kinship and Social Organization.* New York: Macmillan.
Bulatao, R. A.
   1981   "Values and Disvalues of Children in Successive Childbearing De-
      cisions." *Demography* 18 : 1–25.
Caldwell, John
   1982   *The Theory of Fertility Decline.* New York: Academic Press.
Camín, Héctor Aguilar
   1986   *Morir en el golfo.* Mexico City: Oceano.
Caminos, Horacio, John F. C. Turner, and John A. Steffian
   1969   *Urban Dwelling Environments: An Elementary Survey of Settle-
      ments for the Study of Design Determinants.* Cambridge: MIT Press.
Carpizo, Jorge
   1978   *El presidencialismo mexicano.* Mexico City: Siglo Veintiuno.
Chant, Sylvia
   1984a   "Las Olvidadas: A Study of Women, Housing, and Family Struc-
      ture in Querétaro, Mexico." Ph.D. diss., University of London.
   1984b   "Household Labour and Self-Help Housing in Querétaro, Mex-
      ico." *Boletín de Estudios Latinoamericanos y del Caribe* 37 : 45–68.
   1985   "Single Parent Families: Choice or Constraint?" *Development
      and Change* 16 : 636–656.
   1988   "Mitos y realidades de la formación de las familias encabezadas
      por mujeres: El caso de Querétaro, México." In *Mujeres y sociedad: Sa-
      lario, hogar, y acción social en el occidente de México,* ed. Luisa Gaba-
      yet, Patricia García, Mercedes González de la Rocha, Sylvia Lailson,
      and Agustín Escobar. Guadalajara: CIESAS and El Colegio de Jalisco.
Chenery, Hollis, S. Montek, C. L. G. Bel, J. H. Duloy, and Richard Jolly
   1974   *Redistribution with Growth.* Published for the World Bank and
      the Institute of Development Studies, University of Sussex. London:
      Oxford University Press.
Conroy, Michael E., Mario Coría Salas, and Felipe Vilá
   1980   *Socioeconomic Incentives for Migration from Mexico to the U.S.:
      Magnitude, Recent Changes, and Policy Implications.* Austin: Insti-
      tute for Latin American Studies in collaboration with the Instituto Po-
      litécnico Nacional de Mexico.

Davin, Anna
1984 "Working or Helping? London Working-Class Children in the Domestic Economy." In *Households and the World Economy*, ed. Joan Smith, Immanuel Wallerstein, and Hans Dieter Evans. Beverly Hills: Sage.

de Barbieri, Teresita, and Orlandina de Oliveira
1987 *La presencia de las mujeres en América Latina en una década de crisis*. CIPAF. Santo Domingo, Dominican Republic: Ediciones Búho.

de Beauvoir, Simone
1972 *The Second Sex*. London: Penguin.

de Janvry, Alain
1981 *The Agrarian Question and Reformism in Latin America*. Baltimore: Johns Hopkins University Press.

de la Peña, Guillermo
1986 *Cambio regional, mercado de trabajo y vida obrera en Jalisco*. Guadalajara: Colegio de Jalisco.

Duque, Joaquín, and Ernesto Pastrana
1973 *Las estrategias de superviviencia económica de las unidades familiares del sector popular urbano: Una investigación exploratoria*. Santiago, Chile: PROELCE.

Easterlin, Richard
1968 *Population, Labor Force, and Long Swings in Economic Growth: The American Experience*. NBER General Series No. 68. New York: Columbia University Press.
1973 "Relative Economic Status and the American Fertility Swing." In *Family Economic Behavior*, ed. Eleanor Sheldon. Philadelphia: Lippincott.
1978 "The Economics and Sociology of Fertility: A Synthesis." In *Historical Studies of Changing Fertility*, ed. Charles Tilly. Princeton: Princeton University Press.

Eckstein, Susan
1977 *The Poverty of Revolution: The State and the Urban Poor in Mexico*. Princeton: Princeton University Press.

El Guindi, Fadwa
1977 *Religion in Culture*. Dubuque, Iowa: W. C. Brown.
1986 *The Myth of Ritual: A Native's Ethnography of Zapotec Life-Crisis Rituals*. Tucson: University of Arizona Press.

El Guindi, Fadwa, and Henry A. Selby
1975 "Dialectics in Zapotec Thinking." In *Meaning in Anthropology*, ed. Keith Basso and Henry A. Selby. Albuquerque: University of New Mexico Press.

Engels, Friedrich
1952 *The Condition of the Working Class in England in 1844*. London: Penguin. Originally published in 1892.

Escobar, Agustín
1986 *Con el sudor de tu frente: Mercado de trabajo y clase obrera en Guadalajara*. Guadalajara: El Colegio de Jalisco.

Finsten, Laura, and Arthur D. Murphy
    1988   "Domestic Space in Government Housing in Oaxaca, Mexico."
    Presented at the annual meeting of the Society for Applied Anthropol-
    ogy, Tampa, Florida. MS available from Department of Anthropology-
    Sociology, Baylor University, Waco, Texas.
Gabayet, Luisa
    1988   *Obreros somos: Diferenciación social y formación de la clase
    obrera en Jalisco.* Guadalajara: CIESAS y El Colegio de Jalisco.
García, Brígida, Humberto Muñoz, and Orlandina Oliveira
    1982   *Hogares y trabajadores en la Ciudad de México.* Mexico City: El
    Colegio de Mexico/UNAM.
García, Brígida, and Orlandina Oliveira
    1979   "Una caracterización sociodemográfica de las unidades domésticas
    en la ciudad de México." *Demografía y Economía* 13 : 1 – 18.
Geoghegan, William
    1969   "Decision-Making and Residence on Tagtabon Island." Working
    Paper No. 17. Language-Behavior Research Laboratory. Berkeley: Uni-
    versity of California.
Glick, Paul
    1977   "Updating the Life-Cycle of the Family." *Journal of Marriage and
    the Family* 1977 : 5 – 13.
González de la Rocha, Mercedes
    1984   "Domestic Organization and Reproduction of Low-Income House-
    holds: The Case of Guadalajara." Ph.D. diss., University of Manchester.
    1986   *Los recursos de la pobreza: Familias de bajos ingresos en Guada-
    lajara.* Guadalajara: El Colegio de Jalisco and CIESAS.
González de la Rocha, Mercedes, and Agustín Escobar
    In press   "Unidad doméstica y taller: ¿Reductos de resistencia a la cen-
    tralización?" In *Hierarchy and Centralization in Mexico,* ed. Harley
    Browning, Guillermo de la Peña, and Bryan Roberts. Guadalajara: El
    Colegio de Jalisco and CIESAS.
González Casanova, Pablo
    1970   *Democracy in Mexico.* New York: Oxford University Press.
Graedon, Teresa
    1976   "Health and Nutritional Status in an Urban Community in South
    Mexico." Ph.D. diss., University of Michigan.
Grindle, Merilee
    1977   *Bureaucrats, Politicians, and Peasants in Mexico: A Case Study
    in Public Policy.* Berkeley and Los Angeles: University of California
    Press.
Hackenberg, Robert, Arthur D. Murphy, and Henry A. Selby
    1984   "The Urban Household in Dependent Development." In *House-
    holds: Comparative and Historical Studies of the Domestic Group,* ed.
    Robert M. Netting, Richard Wilk, and Eric J. Arnould. Berkeley: Uni-
    versity of California Press.

Harris, Olivia
  1986 "La unidad doméstica como una unidad natural." *Nueva Antropologia* 30: 199–222.
Harrison, Bennett, and Andrew Sum
  1979 "The Theory of 'Dual' or Segmented Labor Markets." *Journal of Economic Issues* 13(3): 687–706.
Hellman, Judith
  1983 *Mexico in Crisis*, 2d ed. New York: Holmes and Meier.
Hirshleifer, J.
  1977 "Shakespeare vs. Becker on Altruism: The Importance of Having the Last Word." *Journal of Economic Literature* 15(2): 500–502.
Hotchkiss, John
  1968 "Children and Conduct in a Ladino Community in Chiapas." *American Anthropologist* 69: 711–718.
Kappel, Wayne
  1976 "Alternative Adaptive Strategies in Three Mexican Towns." Ph.D. diss., University of Arizona.
Kapuscinksi, Ryszard
  1986 *Shah of Shahs*. New York: Vintage Books.
Kemper, Robert Van
  1977 *Migration and Adaptation: Tzintzuntzan Peasants in Mexico City.* Beverly Hills: Sage.
  1981 "Obstacles and Opportunities: Household Economics of Tzintzuntzan Migrants in Mexico City." *Urban Anthropology* 10(3): 212–229.
Kerr, Clark
  1950 "Can Capitalism Dispense with Free Markets?" *American Economic Review* 40(2): 278–291.
  1954 "The Balkanization of Labor Markets." In *Labor Mobility and Economic Opportunity*, ed. E. Wight Bakke. New York: Columbia University Press.
Kim, Myung-Hye
  1987 "Female Labor Force Participation and Household Reproduction in Mexico." Ph.D. diss., University of Texas at Austin.
Lailson, Silvia
  1985 "De mercaderes a industriales." In *Guadalajara, la gran ciudad de la industria pequeña*, ed. P. Arias. Zamora: El Colegio de Jalisco.
Laslett, Peter
  1983 *The World We Have Lost Further Explored.* London: Methuen.
Leeds, Anthony
  1973 "Political, Economic, and Social Effects of Producer and Consumer Orientations toward Housing in Brazil and Peru: A Systems Analysis." In *Latin American Urban Research*, vol. 3, ed. Francine Rabinowitz and Felicity Trueblood. Beverly Hills: Sage.
  1977 "Mythos and Pathos: Some Unpleasantries on Peasantries." In *Peasant Livelihood*, ed. Rhoda Halperin and James Dow. New York: St. Martin's Press.

Lewis, Oscar
  1951   *Life in a Mexican Village: Tepoztlán Revisited.* Urbana: University of Illinois Press.
  1963   *The Children of Sánchez.* New York: Random House.
Liebenstein, Harvey
  1976   *Beyond Economic Man.* Cambridge: Harvard University Press.
Lesourne, J.
  1977   *A Theory of the Individual for Economic Analysis.* New York: North-Holland.
Little, Bert
  1983   "Sibling Similarity and Growth Status and Rate among School Children in a Rural Zapotec Community in the Valley of Oaxaca, Mexico." Ph.D. diss., University of Texas at Austin.
Logan, Kathleen
  1981   "Getting By on Less: Economic Strategies of Lower Income Households in Guadalajara." *Urban Anthropology* 10(3):231–246.
Lomnitz, Larissa
  1977   *Networks and Marginality.* New York: Academic Press.
  1987   *A Mexican Elite Family, 1820–1980: Kinship, Class, and Culture.* Princeton: Princeton University Press.
Lomnitz-Adler, Claudio
  1982   *Evolución de una sociedad rural.* Mexico City: SEP/Fondo de Cultura Económica.
Lorenzen, Stephen A.
  1986   "Employment, Earnings, and Consumption Strategies in Urban Mexico." Ph.D. diss., University of Texas at Austin.
Malina, Robert M., and John H. Himes
  1978   "Patterns of Childhood Mortality and Growth Status of a Rural Zapotec Community." *Annals of Human Biology* 5:517–531.
Malina, Robert M., Henry A. Selby, Wendy L. Aronson, and Peter H. Buschang
  1980a   "Re-examination of the Age at Menarche in Oaxaca, Mexico." *Annals of Human Biology* 7:281–286.
Malina, Robert M., Henry A. Selby, Wendy L. Aronson, Peter H. Buschang, and Cameron Chumlea
  1980b   "Growth Status of Schoolchildren in a Rural Zapotec Community in the Valley of Oaxaca, Mexico, in 1968 and 1978." *Annals of Human Biology* 7:367–374.
Mazumdar, Dipak
  1976   "The Urban Informal Sector." *World Development* 4(8):655–679.
  1981   *The Urban Labor Market and Income Distribution: A Study of Malasia.* New York: Oxford University Press.
Mesa-Lago, Carmelo
  1985   *The Crisis of Social Security and Health Care: Latin American Experiences and Lessons.* Pittsburgh: University of Pittsburgh Press.

Mincer, Jacob
   1962 "Labor Force Participation of Married Women." In *Aspects of Labor Economics*. Princeton: Princeton University Press.
Moynihan, D. Patrick
   1986 *Family and Nation.* New York: Harcourt Brace Jovanovich.
Murphy, Arthur D.
   1973 "A Quantified Model of Goals and Values in Coquito Sector, San Juan, Oaxaca, Mexico." Master's thesis, University of Chicago.
   1979 "Urbanization, Development, and Household Adaptive Strategies in Oaxaca, a Secondary City of Mexico." Ph.D. diss., Temple University.
Murphy, Arthur D., and Henry A. Selby
   1981 "A Comparison of Household Income and Budgetary Patterns in Four Mexican Cities." *Urban Anthropology* 10(3): 247–267.
Murphy, Arthur D., and Alex Stepick
   In press *Adaptation and Inequality in Oaxaca: Political Economy and Cultural Ecology in an Intermediate Mexican City.* Philadelphia: Temple University Press.
Musgrove, Philip
   1978 *Consumer Behavior in Latin America.* Washington, D.C.: Brookings Institution.
   1980 "Household Size and Composition, Employment, and Poverty in Urban Latin America." *Economic Development and Cultural Change* 28(2): 249–266.
Namboodiri, N. K.
   1974 "Which Couples at Given Parities Expect to Have Additional Births? An Exercise in Discriminant Analysis." *Demography* 11: 45–56.
Nerlove, Marc
   1974 "Toward a New Theory of Population and Economic Growth." In *Economics of the Family*, ed. Theodore W. Schultz. Chicago: University of Chicago Press.
Nutini, Hugo, and John Roberts
   N.d. "The Wages of Conquest: Origin, Development, Decline, and Expressive Persistence of the Mexican Aristocracy." MS., University of Pittsburgh.
Ojeda, Norma
   1986 "Family Life Cycle and Social Class in Mexico." Ph.D. diss., University of Texas at Austin.
Park, C. B.
   1978 "The Fourth Korean Child: The Effects of Son Preference on Subsequent Fertility." *Journal of Biosocial Science* 10: 95–106.
Paz, Octavio
   1961 *The Labyrinth of Solitude.* New York: Grove Press.
   1985 *The Other Mexico.* New York: Grove Press.
Pindyck, Robert S., and Daniel L. Rubinfield
   1981 *Econometric Models and Economic Forecasts.* 2d ed. New York: McGraw-Hill

Piore, Michael
　1979　*Birds of Passage: Migrant Labor and Industrial Societies.* Cambridge: Cambridge University Press.
Program of Research on Population in Latin America (PISPAL)
　1978　*Lineas prioritarias para la III ͣ fase de la investigación.* Mexico City: PISPAL.
Poniatowska, Elena
　1980　*Fuerte es el silencio.* Mexico City: Ediciones Era.
Portes, Alejandro, and John Walton
　1981　*Labor, Class, and the International System.* New York: Academic Press.
Ramírez, Arnaldo
　1979　*El regreso de Chin-Chin, El Teporocho.* Mexico City: Grijalbo.
Ramos, Samuel
　1934　*El perfil del hombre y la cultura en México.* Mexico City: Universidad Nacional Autónoma de México.
Riding, Alan
　1985　*Distant Neighbors: Portrait of the Mexicans.* New York: Knopf.
Rodríguez, Daniel
　1981　"Discusiones en torno al concepto de estrategias de supervivencia." *Demografía y Economía* 15(2):238–251.
Rosenzweig, Marc R.
　1973　"The Economic Determinants of Population Change in the Rural and Urban Sectors of the United States." Ph.D. diss., Columbia University.
　1977　"The Demand for Children in Farm Households." *Journal of Political Economy* 85:123–146.
Royce, Anya
　1981　"Isthmus Zapotec Households: Economic Responses to Scarcity and Abundance." *Urban Anthropology* 10(3):269–286.
Rubin-Kurtzman, Jane
　1987　*The Socioeconomic Determinants of Fertility in Mexico: Changing Perspectives.* Monograph 23. San Diego: Center for U.S.-Mexican Studies, University of California.
Sanderson, Warren
　1976　"On Two Schools of Economic Fertility." *Population and Development Review* 76:469–477.
Santamaría, Francisco J.
　1959　*Diccionario de mejicanismos.* Mexico City: Porrua.
Sawhill, Isabel
　1977　"Economic Perspectives on the Family." *Daedelus* 106:116–125.
Scherer, Julio
　1986　*Los presidentes.* Mexico City: Grijalbo.
Schlagheck, James L.
　1980　*The Political, Economic, and Labor Climate in Mexico.* Multina-

tional Industrial Relations Series No. 4. Philadelphia: Wharton School, University of Pennsylvania.

Schmink, Marianne
    1984 "Household Economic Strategies." *Latin American Research Review* 19:87–100.

Schneider, David M.
    1968 *American Kinship: A Cultural Account.* New York: Holt, Rinehart and Winston

Scott, Ian
    1982 *Urban Space and Social Development in Mexico.* Baltimore: Johns Hopkins University Press.

Secretaría de Programación y Presupuesto (SPP)
    1980 *Las actividades económicas en México.* 3 vols. Mexico City: SPP.
    1986 *Diez años de indicadores económicos y sociales de México.* Mexico City: SPP.

Selby, Henry A.
    1971 "Social Organization." In *Biennial Review of Anthropology,* ed. Bernard J. Siegel, Alan Beals, and Steven A. Tyler. Stanford: Stanford University Press.
    1974 *Zapotec Deviance.* Austin: University of Texas Press.

Selby, Henry A., and Lucy R. Garretson
    1981 *Cultural Anthropology.* Dubuque, Iowa: William C. Brown.

Selby, Henry A., and Gary Hendrix
    1976 "Policy Planning and Poverty: Notes on a Mexican Case." In *Anthropology and the Public Interest,* ed. Peggy R. Sanday. New York: Academic Press.

Selby, Henry A., and Arthur D. Murphy
    1979 "The City of Oaxaca." Project Number 931-0003.19. Office of Urban Development of the Agency for International Development. Washington, D.C. MS available from the University of Texas at Austin.
    1982 *The Mexican Urban Household and the Decision to Migrate to the United States.* Occasional Paper in Studies in Social Change. Philadelphia: Institute for the Study of Human Issues.

Simon, Herbert A.
    1978 "Rationality as Process and as Product of Thought." *American Economic Review* 68(2):1–16.

Simon, J. L.
    1975 "The Effects of Income and Education on Successive Births." *Demography* 12:259–274.

Smith, Adam
    1976 *The Theory of Moral Sentiments.* Indianapolis: Bobbs-Merrill. Originally published in 1761.

Smith, Peter H.
    1979 *Labyrinths of Power: Political Recruitment in Twentieth-Century Mexico.* Princeton: Princeton University Press.

Stepick, Alex
 1974 "Values and Decision-Making in Migration." Ph.D. diss., University of California, Irvine.
Stepick, Alex, and Gary G. Hendrix
 1974 *Predicting Behavior from Values.* Social Science Working Paper 46. Irvine: University of California.
Stepick, Alex, and Arthur D. Murphy
 1977 "Housing, Household Economics, and Government Intervention among the Urban Poor." Presented at the annual meeting of the American Anthropological Association, Houston.
Stern, Claudio
 1973 *Las regiones de México y sus niveles de desarrollo económico.* Jornadas 72. Mexico City: Colegio de México.
Torrado, Susan
 1978 *Sobre los conceptos 'estrategias familiares de vida' y 'proceso de reproducción de la fuerza de trabajo': Notas teórico-metodológicas.* Buenos Aires: Centro de Estudios Urbanos y Regionales.
 1981 "Estrategias familiares de vida." *Demografía y Economía* 15 : 204–233.
 1982 *El enfoque de las estrategias familiares de vida en América Latina: Orientaciones teórico-metodológicas.* Vol. 2. Buenos Aires: Centro de Estudios Urbanos y Regionales.
Unikel, Luis
 1976 *El desarrollo urbano de México.* Mexico City: Colegio de México.
Uzzell, J. Douglas
 1972 "Bound for Places I'm Not Known to: Adaptation and Residence in Four Irregular Settlements in Lima, Peru." Ph.D. diss., University of Texas at Austin.
 1974 "Cholos and Bureaus in Lima: Case History and Analysis." *International Journal of Comparative Sociology* 15(3, 4): 54–62.
 1980 "Mixed Strategies and the Informal Sector: Three Faces of Reserve Labor." *Human Organization* 39(1): 40–49.
 1987 "A Hometown Mass Transit System in Lima, Peru: A Case of Generative Planning." *City and Society* 1: 6–34.
Valdés, Ximena, and Miguel Acuña
 1981 "Precisiones metodológicas sobre las 'Estrategias de sobrevivencia.'" *Demografía y Economía* 15(2): 234–237.
Valenzuela, Liliana
 1984 "Neza." *UTMOST Magazine.* Spring.
Vélez-Ibáñez, Carlos
 1983 *Rituals of Marginality.* Berkeley: University of California Press.
Welti Chanes, Carlos
 1983 "Ocupación y fecundidad." In *La fecundidad rural en México,* ed. Raúl Benítez and Julieta Quilodrán. Mexico City: Colegio de México/UNAM.

Willis, Paul
1979 *Learning to Labour.* New York: Columbia University Press.
Wilson, William J.
1987 *The Truly Disadvantaged.* Chicago: University of Chicago Press.
Yanagisako, Sylvia
1979 "Social Organization." In *Biennial Review of Anthropology,* ed. Bernard J. Siegel et al. Stanford: Stanford University Press.

Index